Paul Langley.
Tel 38606 1J1

What do you know about The Earth?

What do you know about

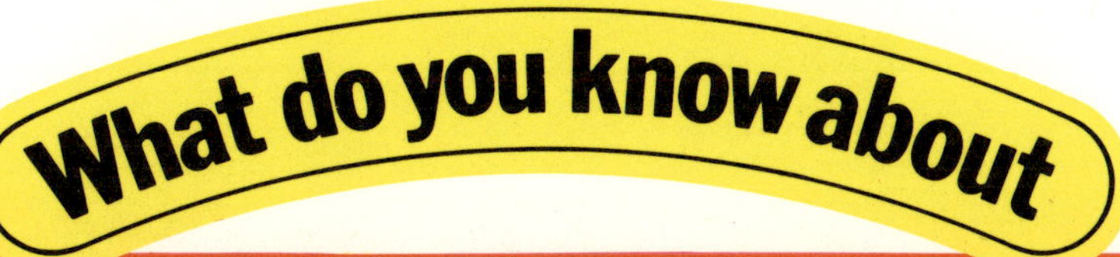

The Earth?

Neil Curtis

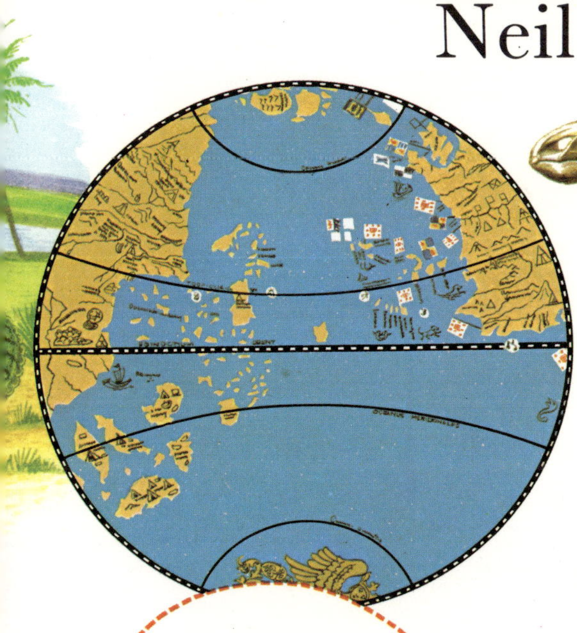

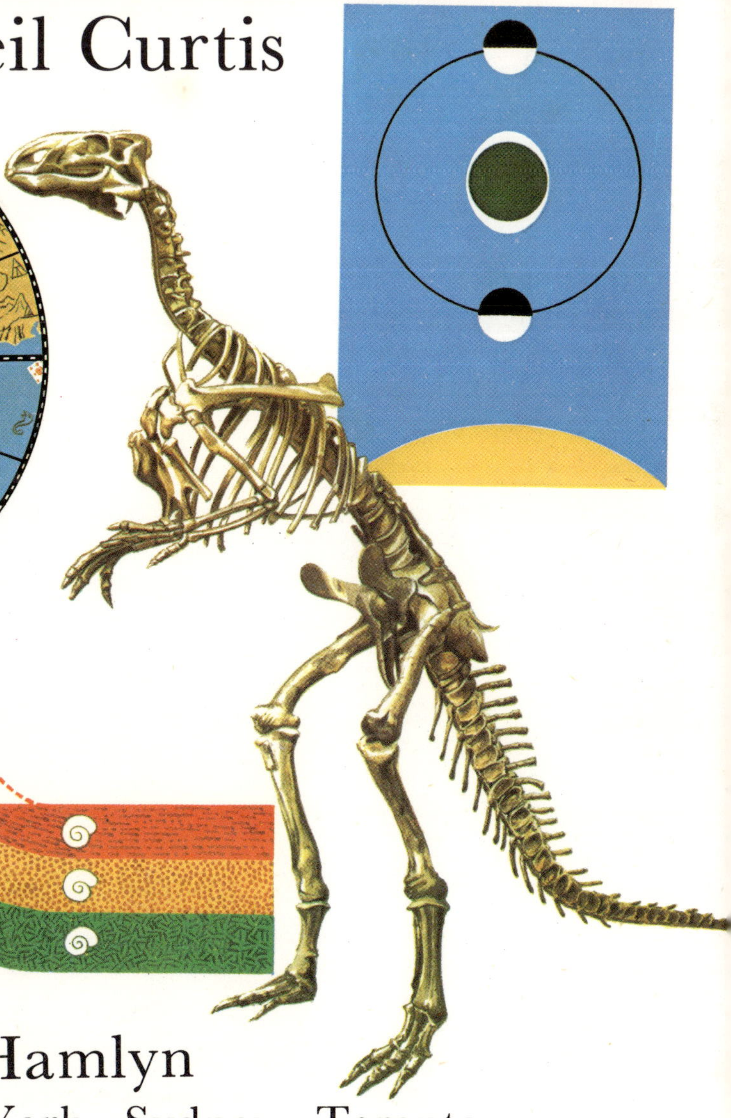

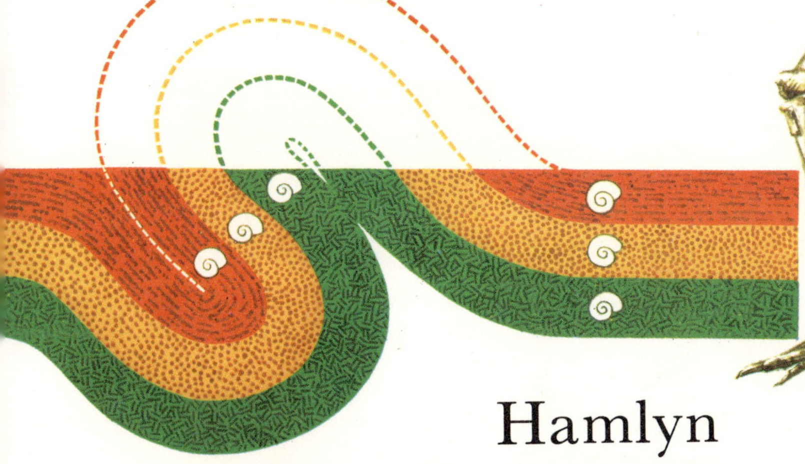

Hamlyn
London · New York · Sydney · Toronto

Contents

Metric measurements have been used in this book except in those cases where Imperial units are still in common usage.

Published 1976 by
The Hamlyn Publishing Group Limited
London · New York · Sydney · Toronto
Astronaut House, Feltham, Middlesex, England

ISBN 0 600 39334 8

Phototypeset by Tradespools Ltd, Frome, Somerset
Printed in Hong Kong by Leefung-Asco Printers Limited

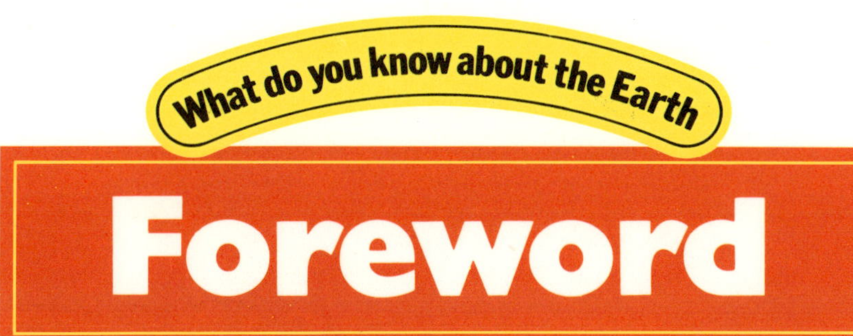

The Earth is the planet on which we live. It provides us with the air we breathe, the food and water we eat and drink, all the raw materials consumed daily by the world's industries, and the very ground we stand on. It is all too easy to take our world for granted and to forget the vast forces locked within the bowels of the planet, or the constantly changing face of the Earth, or the web of living things upon which we rely so completely.

But many people have long been fascinated by the processes going on in and around the Earth, processes which may act so slowly that it would take hundreds of lifetimes to be aware of any change. It is by looking at the record of the rocks and the once-living creatures contained within them that scientists have been able to work out a history of the intriguing development of the Earth, its nearest planetary neighbours, its atmosphere and weather patterns, and the living creatures that dwell upon its ever-changing surface.

In recent years many of the Earth's secrets have been unravelled but there are many, many more questions that remain unsolved. In this book, I have attempted to answer a selection of the huge array of questions concerning this planet in an effort to present a picture of its evolution and workings, hoping that you will then want to know more about the world on which we live.

The author would particularly like to thank Barbara James for her patient help during the writing of the book and Roger Dixon for his invaluable comments on the text. He would also like to express his thanks especially to Rosalind Carreck, and Paul Welti and his colleagues for their creative efficiency throughout all the stages of production of *What do you know about the Earth?*

Neil Curtis

S
W
E
N

What do you know about

The Earth and its Processes

What is the Earth?

You and I already know a great deal about the Earth, because it is the place on which we live. The Earth is a *planet*; that is, an almost spherical ball of rock, soil, and water, surrounded by an envelope of gases which we call *air* or the *atmosphere*.

The shape of the Earth is more correctly called an *oblate spheroid* which is a complicated way of saying that the planet is slightly flattened at the *north* and *south poles* and bulges slightly at the *equator*. This modification of the spherical shape occurs as a result of the Earth's rotation about an axis through the poles. As the Earth spins, its materials tend to be forced outwards. This force is greater at the equator than at the poles.

The Earth, together with its twin planet, or *satellite*, the Moon, are suspended in space and move in an *orbit* around the fiery ball of the Sun at the rate of one complete circuit every year – to be exact every $365\frac{1}{4}$ days. It is because this period of rotation is not exactly 365 days that every four years we need to have a *leap year* of 366 days. This makes up for the quarter days lost in the normal 365 day year.

We have said that the materials of the planet tend to be flung outwards as the Earth goes round. When you look at

Above:
The Earth supports life in a multitude of forms, organized into communities.

Right:
The planet Earth is a partly solid *sphere* about two-thirds covered by water – the seas and the oceans. It provides us with everything we need to live.

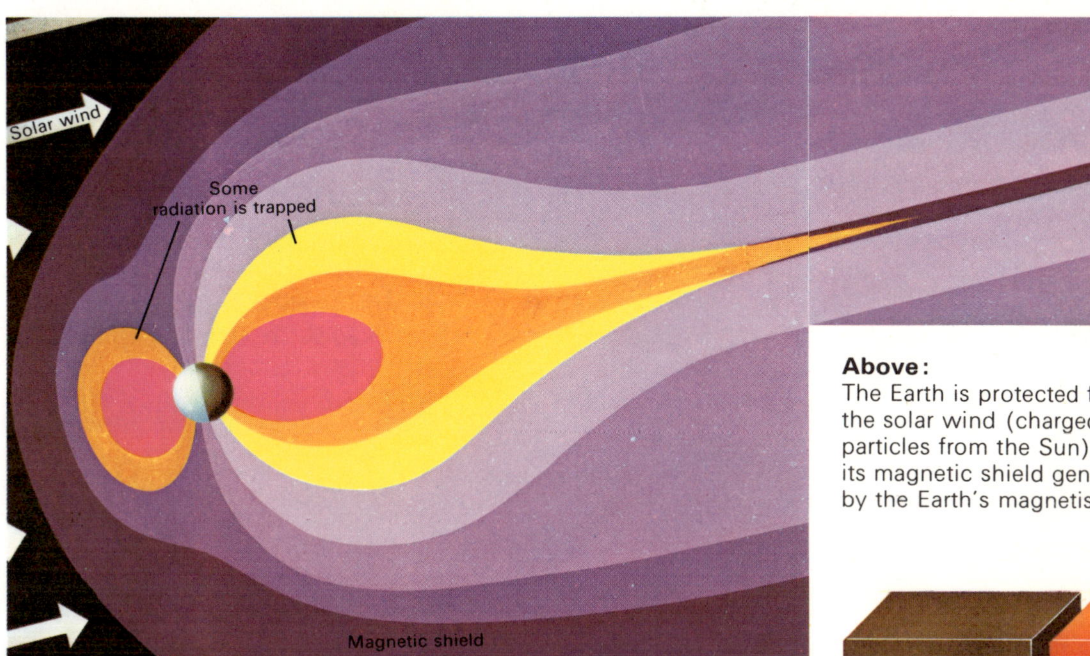

Above:
The Earth is protected from the solar wind (charged particles from the Sun) by its magnetic shield generated by the Earth's magnetism.

the ground beneath your feet or take hold of a piece of rock in a quarry it is hard to imagine that the shape of the whole Earth can change, as you might mould a piece of plasticine. But it is simply a question of the scale of things. The Earth seems to us to be very hard and solid because we are so tiny by comparison. But the forces locked up within the planet are more than a match for the strength of the rocks. It has been said that if we could reduce the Earth to the size of a football and its hardness by an equivalent amount it would have a consistency something like that of toothpaste.

It is upon this base that the whole range of familiar and not so familiar life forms have developed, from the tiny, single-celled amoeba up to and including the most advanced life form the planet has yet known – man. Although man certainly is the most intelligent of Earth's animals, he must remember that he has developed as just one part of the complex network of life and that he is not removed from it.

Perhaps you have heard the expression 'spaceship Earth'. You can understand that if you were isolated in a tiny spaceship for the rest of your life it would be very important to make the best use of all your resources and not to foul the compartment with your waste. This is also true of the Earth, but because it is so big we sometimes tend to forget it.

Above:
The Earth's present atmosphere was formed from original gases at the Earth's beginning, gases from the surface and volcanoes, and oxygen released by plants.

Where is the Earth?

If you have ever been lost in a city or even a large town without the help of a street plan you will have some idea of the problems which confronted scientists of old when they tried to imagine exactly where Earth was in the Universe. You might be able to work out where you were in the city if you could climb on to a very tall building and have a bird's eye view of the city as a whole. In the same way, if you could detach yourself from Earth and look at it from space you would be able to see where it is.

We have explained that the Earth moves around the Sun in the previous question but this was not at all obvious to our ancestors and it has taken many generations of scientific thought to devise an accurate 'street plan' of space which pinpoints the Earth's position correctly.

Like us, our forefathers could look up at the night sky and observe a constantly changing pattern of bright objects including the much larger Moon. We know now that the Earth is just one of nine planets which together with their satellites (moons), a belt of smaller rocky bodies called *asteroids*, some *meteors* which appear to us as shooting stars, and one or two *comets* are all orbiting around the Sun some 140 million kilometres away.

Right and centre: The Earth seems very large indeed to us, but this picture shows the position of our solar system in the known Universe and how tiny even the solar system is.

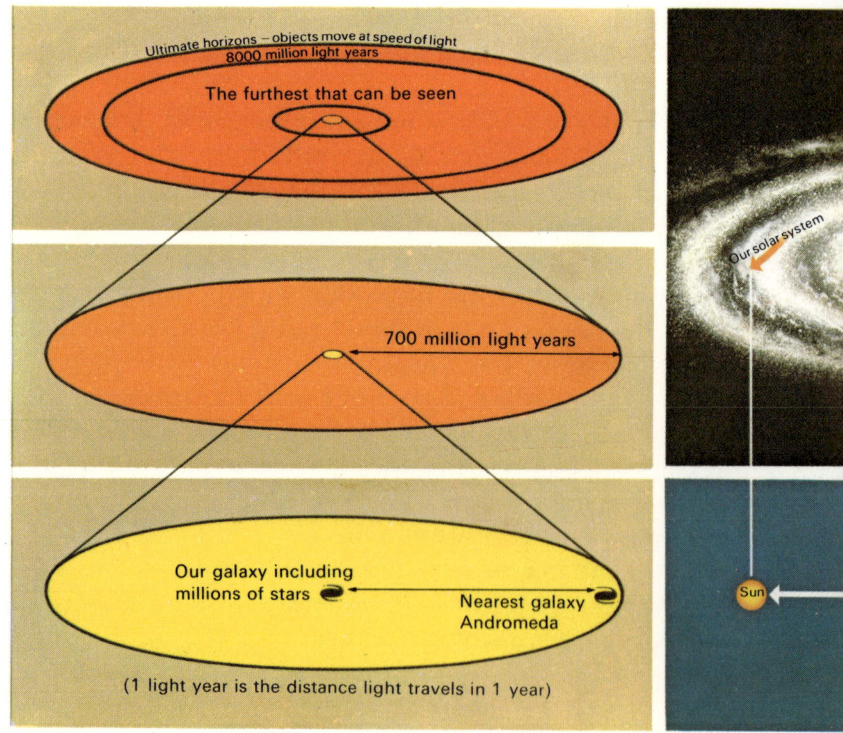

This arrangement of various heavenly bodies is called the *solar system*. But what of all the other myriad points of light which we usually refer to simply as stars? From the vantage point of Earth, these stars appear to be arranged in patterns which have been given names like the *Plough* because of their resemblance in outline to more familiar objects. These patterns are called *constellations* but they may be made up of stars which are many millions of kilometres from each other.

In fact, each one of these pin-pricks of light is a sun very similar to our own except that some of them are younger and cooler and appear to be red in colour and others are hotter, older, and blue. Our solar system is just one of many similar systems which all fit into a larger arrangement called a *galaxy*, and there are millions of other galaxies in the Universe.

It has only been with the coming of the telescope in the 1600s and more recently radio telescopes and satellites that we have been able to establish our 'street plan'. Our ancestors believed that the Earth was at the centre of things. Some ancient civilizations believed that our planet was supported by huge animals such as elephants and whales. They could be forgiven for thinking that the sky was like a gigantic domed roof with the Sun, the other planets and the stars suspended from it.

Below right:
In about AD 150, the Greek astronomer Ptolemy believed that the Earth was at the centre of the solar system and that the planets and the Sun revolved around it.

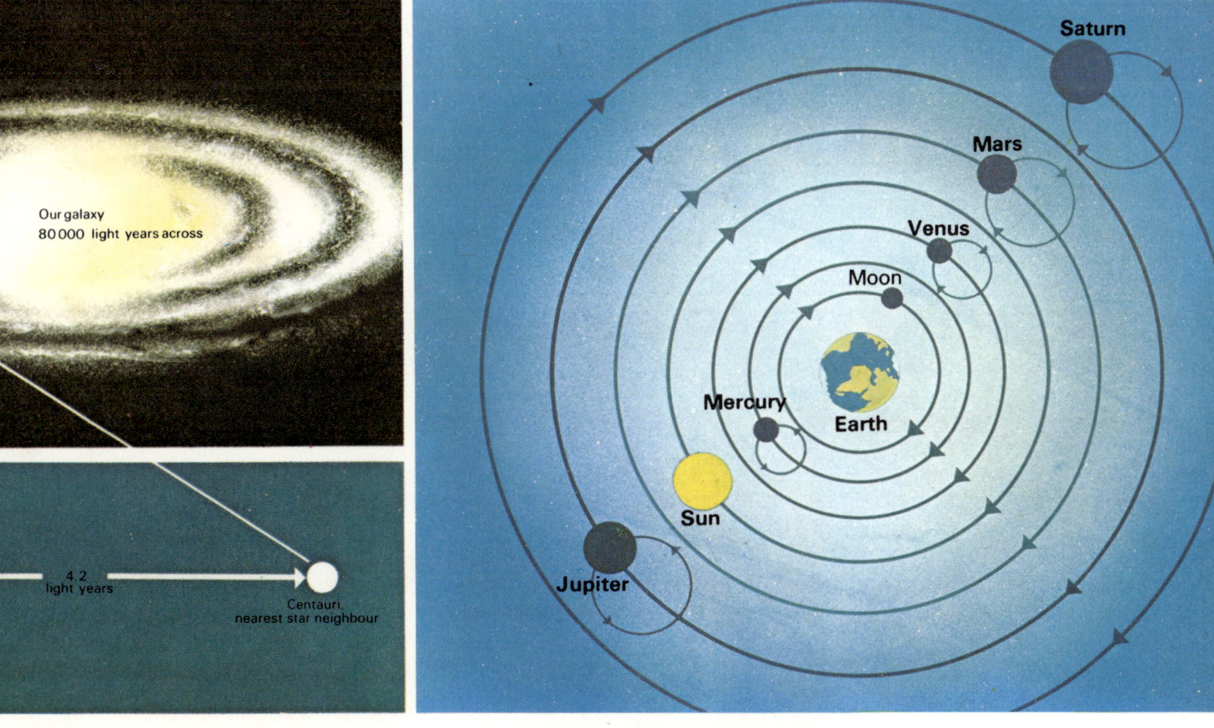

How was the Earth born?

Perhaps you are wondering why we have chosen to ask this question rather than 'How was the Universe born?' Unfortunately, this has been puzzling scientists for many years and has still not been satisfactorily answered.

Many different ideas have been suggested, ranging from the hand of God to the 'Big Bang Theory'. 'Big Bang' suggests that the Universe was formed about 20,000 million years ago by the sudden explosion of a small mass of material which was incredibly heavy. But where did this 'primaeval atom', as it has been called, come from in the first place? It seems unlikely that anyone will ever have the final answer.

On the other hand, once we can assume that the Universe does exist, it becomes a little easier to think about the formation of our own planet because we already have the building materials available.

For many years, the most accepted idea concerning the formation of the Earth was the result of the work of a scientist called Kelvin in 1862. He suggested that the Earth has cooled from a melted, that is, a molten, mass. He put forward this idea partly because it had been proved definitely by that time that the Earth is losing heat, and also because this seemed to provide an explanation for the origin of volcanoes spewing their red hot lava from the interior.

More recently, however, a great deal of evidence has been gathered, indicating that the planet was probably never completely melted. In fact, the most popular theory nowadays is that the Earth, and also the rest of the solar system, were formed from the coming together of a cloud of dust and gas. This was probably a result of the gravitational

Above:
Buffin thought that the Earth was formed hot and that it cooled from the outside towards the centre. This made the surface crack to build mountains and valleys.

Right:
About 450 million years ago (a mere tenth of the age of our planet) the surface of the Earth may have looked like this.

attraction which all objects have for each other. We think this occurred about 6000 million years ago. This means that the Earth probably formed cold rather than as a molten ball.

We know from the evidence of deep mines, however, that the Earth becomes very much hotter towards the centre. Where does the heat come from? Certainly, it must have become hot enough to melt many of the original materials. We think this heat has a twofold origin. When the Sun contracted at the very beginning, its centre became hot. This set off the decay of certain radioactive substances which released a great deal more heat. It seems likely that the first of the Earth's rocks were formed from their molten parent materials about 4500 million years ago.

Left:
This picture illustrates the traditional idea of the creation of the Earth as related in the Old Testament of the Bible.

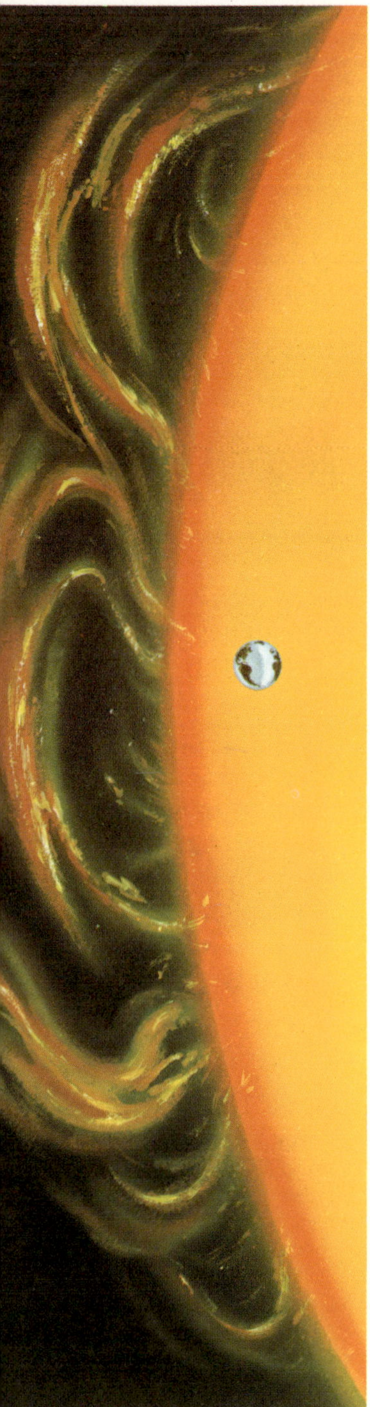

Above:
Look at this illustration and see how tiny the Earth is, compared to the Sun.

How does the Earth keep warm?

This is a question that you have answered for yourself every time that you have been sunbathing. Almost all of the Earth's heat, and this means its energy, must come from the Sun. In addition to this the interior of the Earth has retained much of the heat which was generated at the formation of the Earth. This internal heat came from energy released as the Earth contracted as well as from heat given off by the decay of certain types of radioactive materials. But more of this later. For the moment we are concerned with the Sun. No plants would grow without the Sun, and therefore, no animals and no man. Without the Sun, our planet would be a frozen, barren waste.

What is the Sun?

In an earlier question, we explained that the Sun is a star like many millions of others in the Universe, but why does the Sun or any other star shine and give out its warmth?

The Sun is more than one hundred times larger in diameter than the Earth, and over a million times larger in volume, but its mass is only about 330,000 times more.

You know that the Earth is solid and that gases are usually lighter than solids, so that if you were to guess that the Sun was made up mostly of gas you would be correct. By using an instrument called a *spectrograph*, which carefully examines the kind of light which the Sun is emitting, scientists have

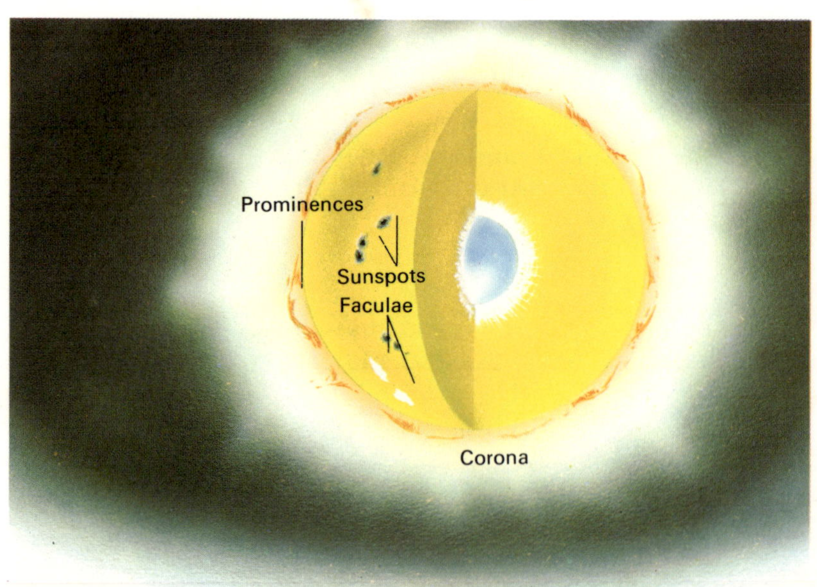

been able to prove that 70 or 80 per cent of the Sun is made up of hydrogen, which is the lightest material that we know of. The rest is mainly another very light gas called helium, from the Greek word *helios* meaning sun because it was first noticed there. You may have come across these two gases because they have been used to fill airships and balloons.

Why is the Sun able to burn for so long without ever seeming to use up its fuel? The process is simple to explain, but cannot successfully be repeated on Earth even in a controlled way.

All elements are made of tiny particles called *atoms*. The atoms of different elements are made up of different numbers of other particles, such as *protons*, *neutrons*, and *electrons*, giving each element its own special properties. In the Sun, atoms of hydrogen are constantly combining by a process called *fusion* into atoms of helium. This happens at a temperature of about fourteen million degrees Centigrade (remember that water boils at only 100°C) and a tremendous amount of heat and light is liberated. There is so much hydrogen fuel that it will last a very long time, particularly because so little hydrogen releases so much heat.

Right:
A view of the Sun showing faculae and sunspots.

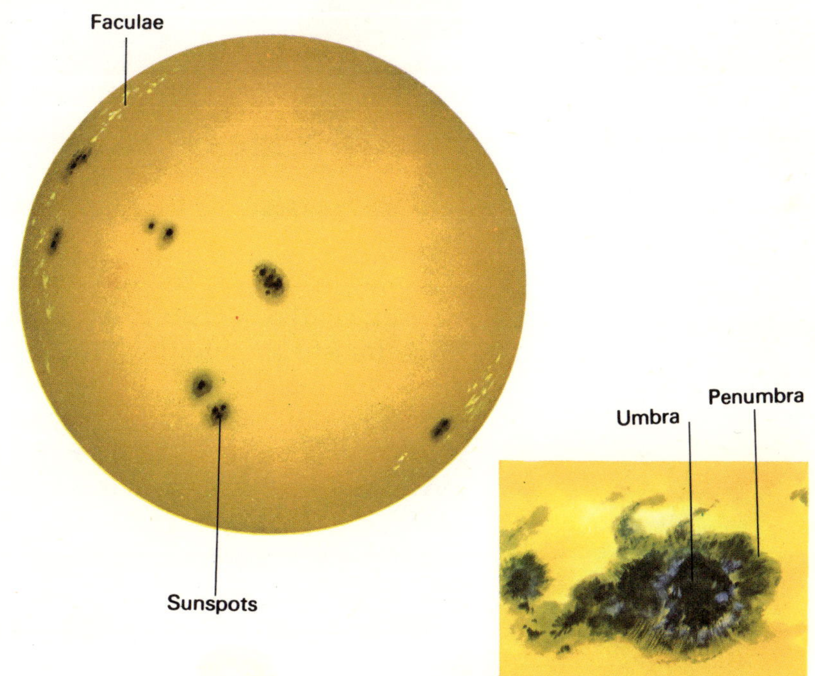

Left:
A section through the Sun showing what the various parts are called. Faculae are bright areas and sunspots are dark spots that both appear on the surface. Prominences are clouds of intensely hot gas which appear to shoot up like flames from the surface of the Sun.

Above:
A close up of a sunspot shows that it has a darker inner part called the umbra and a lighter outer part called the penumbra.

Has there ever been a Man in the Moon?

The legend of the Man in the Moon has grown up for a very obvious reason. If you look up at the silvery disc of a full Moon on a clear night the surface features of our satellite could easily give you the impression of a face. Before the coming of space travel many science fiction writers told of strange creatures which inhabited our neighbour. H. G. Wells wrote a very famous story called *The First Men On The Moon* in which he tells of ant-like beings which he called 'Selenites'. These 'men' lived in the craters and caves of the Moon and would break at the lightest blow of a human fist.

Right:
The American Apollo space programme was started to find out more about the Moon. Here we see a manned spacecraft in orbit around the Moon. The Earth can be seen in the distance.

Today, space probes and manned landings on the Moon have given us new clues to its origin and history. Scientists are now almost sure that there has never been life on the Moon. It was once thought that originally the Earth had no Moon, but as a result of a terrific explosion the Moon was shot out into space from what is now the Pacific ocean. Geologists have long since discarded this idea. It now seems to be more likely that the Moon and the Earth were formed at more or less the same time, contracting from a cloud of dust and gas, as we have explained in a previous question.

If this is true, why do they appear to be so different? The answer is one of chance. When the Moon was born it was too small (it has about one-sixth of the Earth's gravity) to retain an atmosphere or any lighter, life-giving materials like water. This means that the Moon is, and always has been a dry, barren waste without life of any kind. You can see, then, that there never has been a Man in the Moon but there has already been more than one man on the Moon!

Left:
As you can see in this illustration, the surface of the Moon is pitted with many craters.

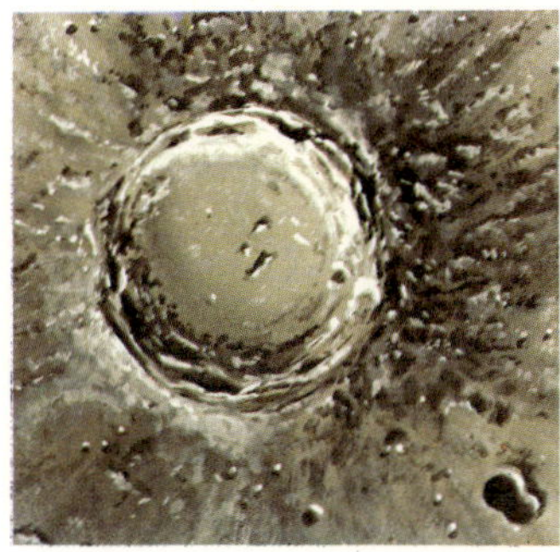

Above:
A close-up of the crater Copernicus.

When will the Earth die?

As you have already seen, the very existence of the Earth, and all the other planets in the solar system, depends upon the Sun. You have seen how the Sun provides enormous amounts of energy by the change of the gas hydrogen into the gas helium. But as with any other 'engine' the fuel, that is the hydrogen, must slowly be used up. When this happens, as has already occurred with millions of other stars, the Sun will collapse and explode. The explosion will totally destroy the Earth and its neighbours. There is no need to worry unnecessarily – it will be at least another 5000 million years before the Sun explodes.

Left:
Galaxies may attract each other by gravitational pull and even collide.

Who has explored the Earth?

Sometimes our planet Earth seems very tiny. We live in the age of jet aircraft that can travel at more than 800 kilometres an hour and space rockets that are very much faster. Even in our daily lives, most of us would not think it unusual to cover a distance of 2000 kilometres to spend a fortnight's holiday. And, of course, except for a few remote parts of South America and New Guinea, the world is well known to us, if not by first hand experience, at least through television programmes, films and books. It has not always been so!

Right:
The so-called 'hairy people', met by explorers in Africa before the birth of Christ, were probably chimpanzees.

Even a hundred years ago, it would have been quite an undertaking to travel by road from Edinburgh to London, riding in a coach drawn by four or six horses which had to be changed at frequent intervals. But most of the world had been explored, even if only by a few hardy men and women. More important, perhaps, was the fact that it was generally accepted that the Earth was a globe and was not flat as had been believed by earlier generations.

It is important to remember that man did not evolve at the same speed all over the world. While some civilizations were recording their history, learning the sciences and making new discoveries others were still in a primitive state. Many of the voyages of exploration came as a result not just of human curiosity but the need to find new trade routes. As horizons were widened the centre of trade shifted from the Mediterranean countries to those facing the Atlantic. Do not forget that the early voyagers had only small, wooden ships and navigated the oceans with the aid of their knowledge of the position of the Sun and the stars.

Right and below:
Viking settlers like Leif Eriksson may have been the first white men to meet North American Indians, but the discovery of America is usually attributed to Christopher Columbus.

As long ago as 600 years before the birth of Christ, kings of Egypt were sending hardy seamen on perilous voyages in an attempt to sail around the coast of Africa. Later, Alexander the Great took his armies across Asia to India, founding many cities on his way. The most famous is Alexandria on the coast of North Africa.

There are many other great names in the history of the discovery of our world. Marco Polo immediately springs to mind with his epic crossing of Asia to reach the court of the Moghul emperor, Kublai Khan at Cambaluc (now Peking). In 1492, Christopher Columbus came across the West Indies by accident in his efforts to discover a new trading route to the East Indies, the source of the profitable spice trade. He is usually attributed with the discovery of America although it had probably been found 400 years earlier by the Viking sailor Bjarni Herjolfsson. Today, the explorer must go much further afield, already to the Moon and perhaps to more distant planets in the future.

How do we travel the Earth?

Being able to live in Britain and holiday in the United States of America is something that has been taken for granted for some years by those who are able to afford an air-ticket. Even the least well-off think nothing of seeing and hearing a jet aircraft rushing at speeds of more than 800 kilometres an hour to the other side of the world. Many families in the industrial countries of western Europe and America own some kind of motor vehicle. Even those of us who do not own a car are able to take advantage of the various forms of public transport. Travelling large distances with ease and comfort is something that has become commonplace.

The ability to travel quickly and safely has meant that people are able to work in the big cities and live in more pleasant surroundings out of town. But it is not only for our own travel that we have come to rely so heavily on transport systems. How would our milk arrive on the doorstep each morning without them? Or how would oranges or more exotic fruit and vegetables be so readily available? Or how would people living in Europe be able to buy Japanese motor cycles or cameras? How would children living in remote districts be able to get to school? How would farmers be able to get their produce to market? The list is almost endless.

But travel and transport of goods has not always been so easy. It is worth remembering that it is little more than twenty years since jet aircraft were a novelty and less than a hundred years since the first motor car, a German Daimler Benz, chugged and coughed its way out of the inventor's

Above:
An open boat powered by oars and a single sail is a far cry from the fast, luxurious, ocean-going liners we are used to.

Right:
An old London bus.

Above:
Modern aircraft are powerful enough to carry cars and passengers.

Above right:
Many people rely on fast electric trains to carry them to and from work.

workshop. Although steam driven railway engines were already quite common in industrialized countries in the late 1800s, it was the realization of the properties of petrol and the invention of the internal combustion engine that made the most impact.

For more than half a century, motor cars and the availability of cheap petrol have changed our way of life in

Right:
A nuclear submarine may remain submerged for two years or more.

industrialized countries. Fast, cheap transport has brought commodities from all over the world to our shops. Recently, however, it has been brought to our attention very dramatically that fossil fuels are fast running out. Scientists are hurriedly trying to invent new forms of power, and in the meantime people are beginning to return to a much older form of transport, the horse.

What are our nearest neighbours like?

In our solar system the three bodies that are nearest to us are our own Moon and the planets Venus and Mars. The formation, structure, and surface of the Moon are more closely linked with Earth and we have looked at these similarities in a previous question.

For the moment then, let us concern ourselves with our neighbouring inner planets. Because these two planets are so close to us (in terms of the size of the solar system) they have been the subject of many science fiction stories. But what are these planets really like? And do they really have strange life forms inhabiting them?

Venus is closer to the Sun than the Earth, and orbits at a distance of about 110 million kilometres from the Sun. Sometimes Venus comes as near as 40 million kilometres to Earth. It does seem unlikely, but a year on the planet Venus would be shorter than a day. This occurs because the planet is spinning on its own axis more slowly than it is orbiting around the Sun. In the case of the Earth, of course, it is the other way round so that there are $365\frac{1}{4}$ of our days in a year.

Venus is about the same size and weight as the Earth but because it is much closer to the Sun and does not have our

Venus

Earth

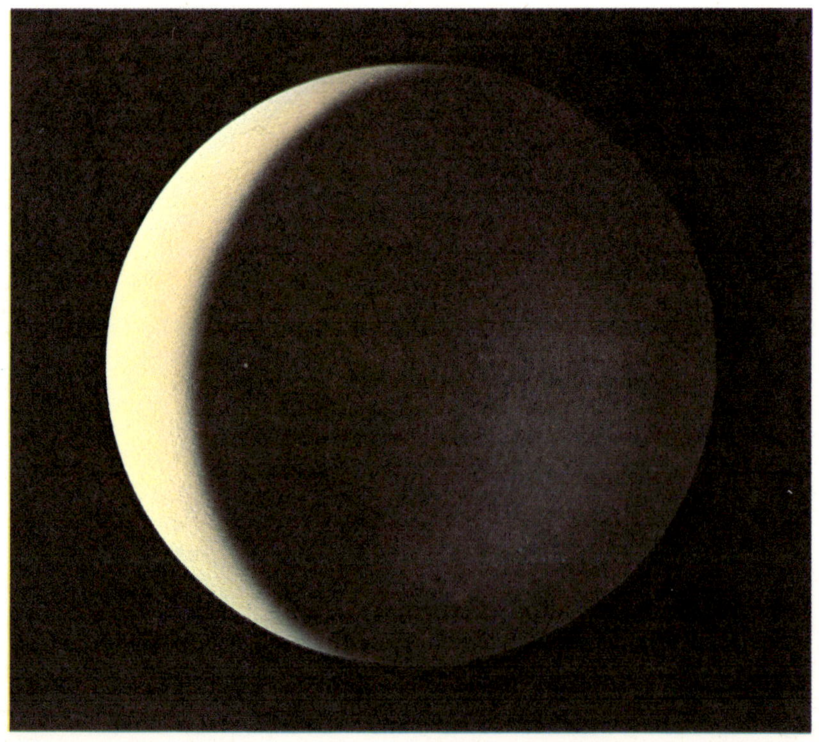

Above and right:
The phases of Venus, that is the ways in which Venus appears to us at different times of the year, are unlike those of the Moon. Venus is the most brilliant of all the planets.

Right:
The two moons or satellites of Mars: Phobos and Deimos.

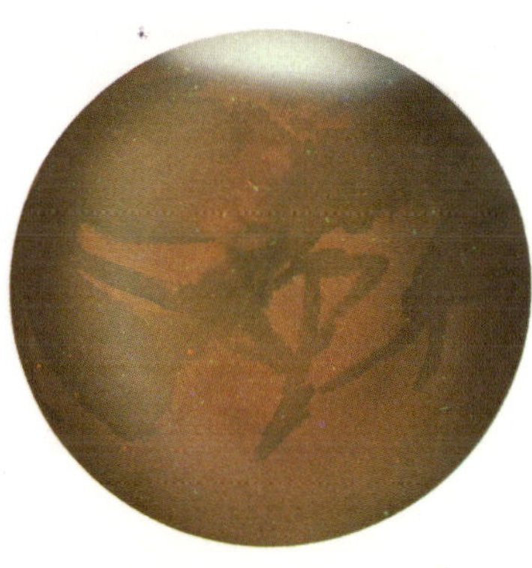

Above:
This picture of Mars shows the north polar icecap in winter.

protecting layer of the gas ozone, it is very much hotter. In fact, the surface temperature has been estimated at about 350°C, or more than three times the boiling temperature of water. Because of the extreme heat, because there does not seem to be a breathable atmosphere apart from a dense covering of carbon dioxide gas, and because there is little evidence of water it would be very surprising if there were any life forms like our own.

Mars has been a particularly favourite haunt of the story-teller's spacemen. H. G. Wells popularized the idea of Martians in his book *The War of the Worlds*. Mars orbits around the Sun at a distance of about 230 million kilometres and is farther away from it than we are. If you look at Mars in the night sky it appears to be red in colour. This reddish tinge seems to be the actual colour of the planet's surface because it is only protected by a thin atmosphere of carbon dioxide.

The surface of Mars is certainly cratered rather like our Moon and also appears to have canal-like scars running across it. It would be very unlikely that any life as we know it could survive there because of the lack of water and free oxygen, and because the temperature may fall low enough during the night to freeze any normal living creature. You can see, then, that Earth is very lucky to have conditions which will support a network of plant and animal life.

Right:
It has been suggested that the Star of Bethlehem was a comet.

Does a comet foretell disaster?

Before we can answer this question, we must look at exactly what a comet is. As you probably know, a comet looks rather like a star with a long, milky 'tail' stretching across the sky. In fact these 'tails' are not tails at all because they always point away from the Sun and do not trail behind.

Right:
The great comet was even visible in daylight in 1910.

Far right:
Brooks's comet last approached the Sun in 1967, but was not visible.

Right:
Halley's comet was last seen in 1910.

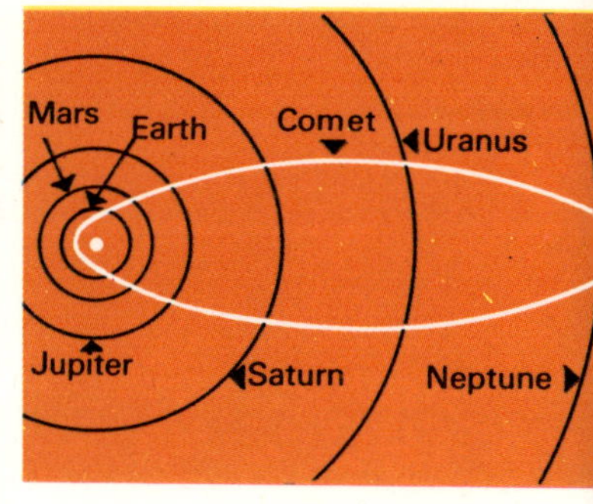

Above:
The orbit of Halley's comet.

A comet seems to be made up of three main parts. The *nucleus* in the centre of the star-like part is composed mainly of ice particles. This is surrounded by a *coma* of much smaller particles and gas. The coma results from the melting of the ice and other substances as the comet approaches the Sun on its extremely stretched out elliptical orbit. The Sun's radiation also tends to force some of these tiny pieces of ice away from the comet and out into space, which explains why the tail is always directed away from the Sun.

A scientist called Halley, after whom a particularly bright comet is named, discovered how comets moved. He also predicted when some comets which had already been seen would return. In fact, Halley's comet is due to be seen by us on Earth in the year 1986.

Why are comets associated with disaster? It is true that a comet appeared before the Norman conquest of Britain in 1066, and that two more comets were seen before the Great Fire of London in 1666 and the Great Plague in the previous year, but in world terms these cannot be regarded as major events. Perhaps it is the result of people's belief that what happens in their own country is of world importance that has led to this idea that comets predict disaster. On the other hand, it has been suggested that the Star of Bethlehem that told of the birth of Christ was in fact a comet. It does seem unlikely, however, that a body of ice and gas moving around the sky in a quite regular orbit can have anything to do with the events here on Earth. This is particularly true of comets because they have such a low density. Although they may have a diameter of as much as 128,000 kilometres they do not have enough gravitational pull to affect the bodies they pass. In fact, their own orbits may be completely changed.

Below:
De Cheseaux's comet has a multiple tail.

How can you look at the stars and planets?

We have already mentioned what an important part the invention of the telescope has played in man's exploration of the Universe. More recently, of course, you have seen the value of radio telescopes, satellites, and even manned spacecraft. But in astronomy (the science of the planets and stars) and in geology (the science of the Earth) there are still plenty of discoveries to be made by the home scientist with the aid of a few readily available instruments.

If you want to look more closely at the stars and planets yourself you will either need a pair of powerful binoculars, or an astronomical telescope. They will give a much greater magnification; that is, they will make the object you are looking at appear much larger.

Above:
A simple astronomical telescope.

Above:
Radio telescopes such as this one at Jodrell Bank with its 75 metre dish are used to track satellites.

Let us look at binoculars first. If your family owns a pair or if you are thinking of buying binoculars, you will find that they are usually described by two numbers and that they come in a variety of prices. The price is dependent upon the quality of the lenses and prisms (these are the pieces that gather the light), on how light in weight they are, and on how much rough handling they will take. Usually it is better to buy the very best pair that you can afford, and you should ask for advice on this. The numbers will appear as 8 × 40

Above:
Binoculars can be used for many hobbies as well as astronomy.

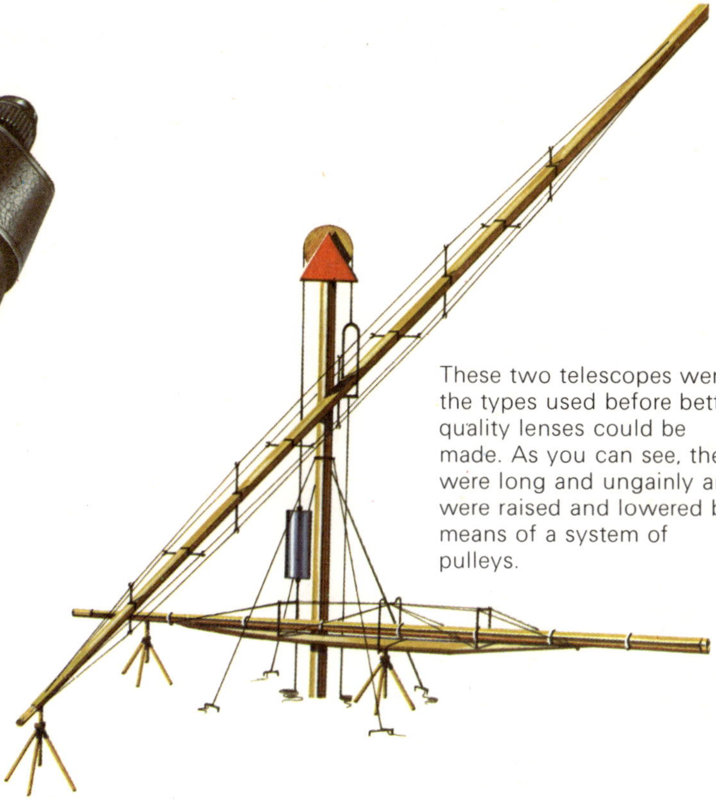

These two telescopes were the types used before better quality lenses could be made. As you can see, they were long and ungainly and were raised and lowered by means of a system of pulleys.

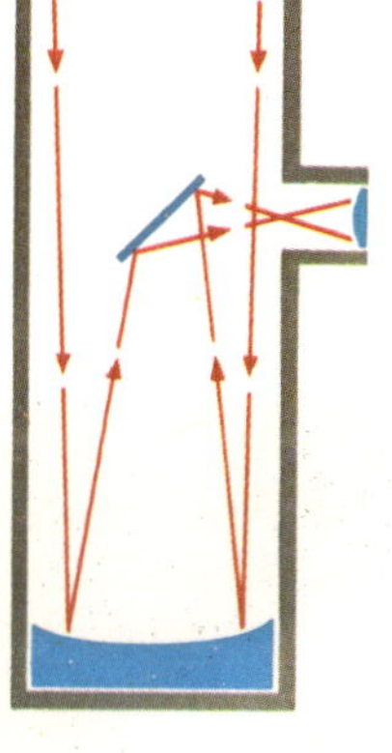

Above:
A reflecting telescope works by gathering light and reflecting it to the eyepiece.

or 7 × 50. The first number is the magnification and the second is the diameter of the largest or object lens measured in millimetres. It is not necessarily better to buy those with the greatest magnification because they cannot usually be used at night and they are very difficult to hold steady without a support. The 7 × 50 binocular is very useful and can also be conveniently used for other hobbies such as birdwatching.

An astronomical telescope is more useful for a closer look at the heavenly bodies. (The difference between an astronomical and other types of telescope is that the image which you see in the astronomical type is upside down and back to front.) Of course, this does not matter when you are looking at something like a star, and it means that an astronomical telescope is generally simpler and less expensive than an equivalent terrestrial telescope (one in which the image is the right way round). Basically, a telescope is an instrument which gathers as much of the light as possible coming from a body. This may be achieved either by a system of lenses, as in a *refracting* telescope or by a large, special type of mirror, as in a *reflecting* telescope. There are many detailed books on this subject that you can find in your local library, and the manufacturers will be only too pleased to help you.

Is there life on other planets?

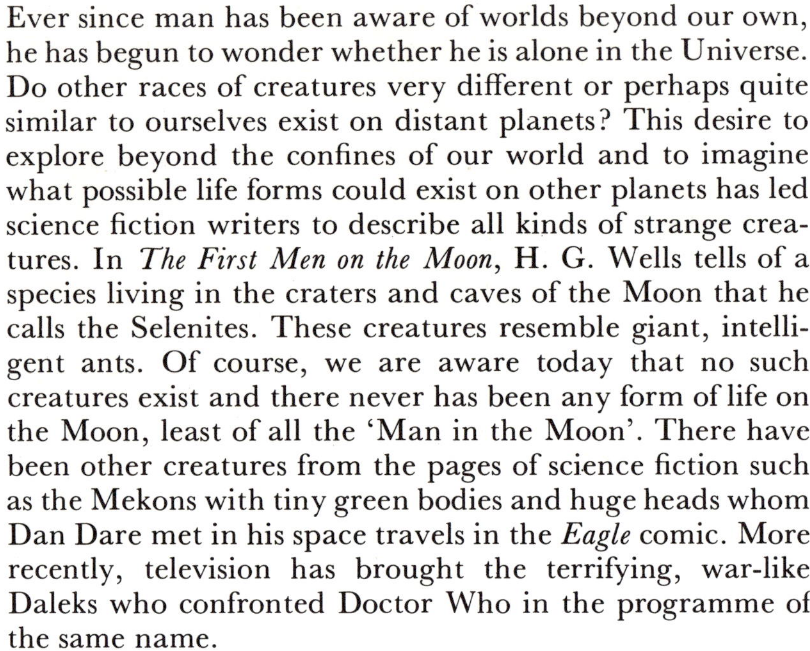

Ever since man has been aware of worlds beyond our own, he has begun to wonder whether he is alone in the Universe. Do other races of creatures very different or perhaps quite similar to ourselves exist on distant planets? This desire to explore beyond the confines of our world and to imagine what possible life forms could exist on other planets has led science fiction writers to describe all kinds of strange creatures. In *The First Men on the Moon*, H. G. Wells tells of a species living in the craters and caves of the Moon that he calls the Selenites. These creatures resemble giant, intelligent ants. Of course, we are aware today that no such creatures exist and there never has been any form of life on the Moon, least of all the 'Man in the Moon'. There have been other creatures from the pages of science fiction such as the Mekons with tiny green bodies and huge heads whom Dan Dare met in his space travels in the *Eagle* comic. More recently, television has brought the terrifying, war-like Daleks who confronted Doctor Who in the programme of the same name.

These are just a few of the imaginary men from outer space, but what are the possibilities of actual life forms? The first point to make is that there is little to be gained from trying to picture weird and wonderful creatures sprouting radio antennae from their heads and breathing liquid ammonia.

Top:
The world's first liquid-fuelled rocket.

Above:
The space craft invented in fiction by Jules Verne.

Right:
Space walks are not just left to the minds of science fiction writers nowadays.

Right:
An artist's impression of the surface of Mars, the imagined home of fictional spacemen.

We should try to confine ourselves to examining whether or not there is any chance of living systems much like our own somewhere else in the Universe. It must also be remembered, however, that a living world elsewhere could be more or less developed than our own. In other words it could still be at the stage of a 'primordial soup' of primitive life forms or thousands of years more advanced than our own world.

Scientists have attempted to calculate how many galaxies there are in the Universe and how many solar systems are contained within these galaxies in an effort to work out how many planets there might be similar to our own. Even the most cautious would probably agree that the number of planets in the Universe similar to our own probably runs into the thousands. This means that the chances are heavily in favour of there being many other worlds just like our own in which there are plants and animals and humans breathing air made up of oxygen, nitrogen, carbon dioxide, and a few other gases.

Are these intelligent creatures trying to make contact with us using radio waves? Some people think so, but of course, they may be thousands of light years away so that by the time we received any message, they could be extinct. The possibility exists that we have already been visited by spacemen and that such wonders of the world as the pyramids of Egypt were constructed with their help. Perhaps soon we shall know more of the creatures from outer space.

What makes the ground tremble?

Perhaps you have seen science fiction films in which great cracks suddenly appear in the ground, and the land seems to roll like the sea. Of course, if you live in Britain or Europe this does not seem very realistic, does it? But, even though

Right:
The San Francisco earthquake in 1906 caused terrible damage, largely due to fires that resulted from fractured gas mains. It is thought that there could be an even more severe tremor in that area at any time along the San Andreas fault. But earthquake waves travelling throughout the Earth can be bent or blocked completely by the different layers. The resulting waves can be detected by seismometers.

these scenes may be exaggerated to make the film more exciting, there are many places around the world, such as Los Angeles in America or Tokyo in Japan, where the ground could tremble at any time, with terrible effects.

This shaking of the Earth is called an *earthquake*. You know that if you throw a stone into a pond, waves spread through the water in all directions from the point where the stone fell. In a similar way, if some rocks in the Earth are somehow disturbed, movements like waves are set up in the ground itself. The land can be felt to shake close to where the rocks were disturbed, and depending on the force of the earthquake, buildings may come tumbling down, water pipes may be broken, roads torn up, and huge fires started by the breaking of gas mains or electricity cables. Last, but certainly not least, thousands of people may lose their lives.

Try placing a brick on a plank of wood and slowly tilt the plank from one end until the brick slips. You will find that you can tilt the plank quite a lot before the brick eventually slips, and when it does it slips suddenly. This is because both the brick and the piece of wood have rough surfaces and tend to grip together. Before the brick can move, this tendency has to be overcome by tilting the plank more and more until the brick suddenly slides. Similarly, inside the Earth, if rocks are pushed more and more they will break, or move very suddenly along existing cracks. Your brick and plank were quite small and you would feel very little when the brick slips, but in the Earth the rock masses are so big that even if they move a little an earthquake is caused.

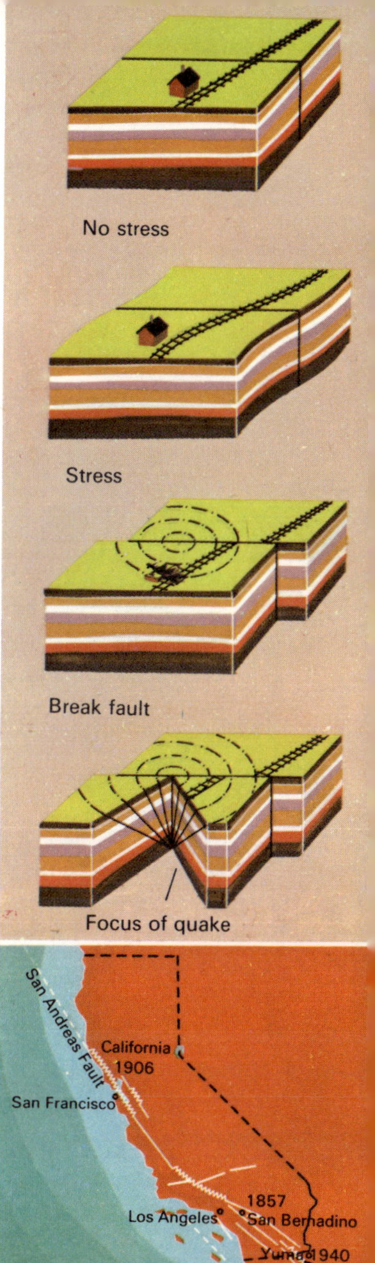

If you were to make the surfaces of your brick and plank more slippery (perhaps by rubbing them with soap) you would find that the plank need not be tilted so much before the brick moved, and that it would slide more slowly. In America, scientists are testing the effects of pumping thousands of gallons of water into the ground in the hope that rocks can be made to move more slowly, and in this way prevent earthquakes.

Apart from the harmful results of earthquakes the ways in which earthquake waves pass through the Earth can tell scientists quite a lot about what the inside of our planet is like. But they have to be able to detect and measure these waves. An instrument which measures earthquakes is called a *seismometer*. Simply, it consists of a masonry column bedded into the solid rock to which is attached an arm loaded with a heavy weight called a *boom*. On the end of the boom is a pen which marks a line on to a piece of paper wrapped round a drum which rotates when the seismometer is in operation. Because of its weight, the boom tends to remain still during an earthquake but the rest of the instrument is shaken and these movements are recorded as a squiggly line on the paper as it rotates.

Top:
This illustration shows how the stress builds up along a fault line, the rocks eventually fracturing, causing an earthquake.

Above:
This map shows the fault systems in California and the epicentres of historic earthquakes in the area.

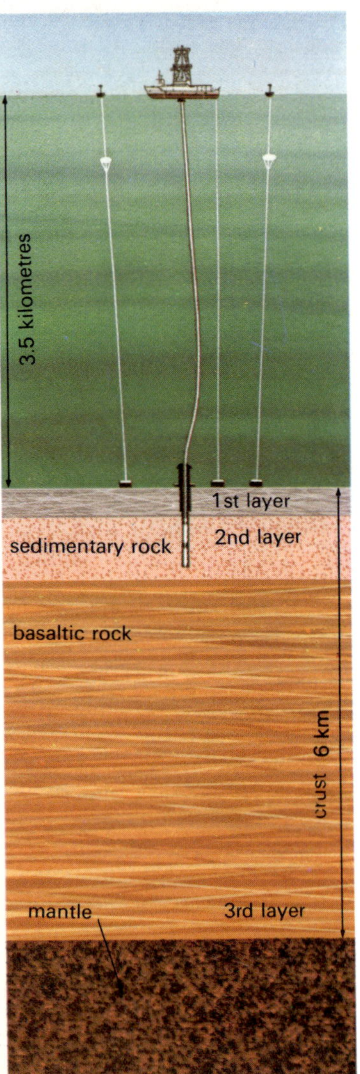

What is the Earth made of?

There are really two questions here. Not only 'What is the Earth made of?' but also 'How do we know what the Earth is made of?'. We can find out about the materials that make up the surface of our planet by simply taking samples of all the rocks and soils which are exposed and carefully examining them. But the Earth measures almost 6600 kilometres from the surface to its centre. The deepest holes which man has been able to drill in his search for oil are a little over seven kilometres and the deepest mines still less. An attempt was made by scientists from the United States of America to drill a hole right through the Earth's crust. They knew that the crust was at its thinnest under the oceans, but they soon discovered the enormous cost and the project was abandoned. This attempt was called the 'Mohole project' in honour of a famous geologist called Mohorovičić. So how do we know what is in the innermost parts of our planet?

Scientists have been able to piece together evidence from a variety of sources to give a picture of the composition of the Earth as a whole but first we should look at some of the more common materials of which the Earth is made.

Let us begin with the air we breathe. It can be shown quite easily that air is mainly a mixture of three gases. There is

Above:
Drilling ships like this one can yield a great deal of information about the composition of the Earth.

Right:
This illustration shows how nitrogen is continuously cycled through animals, plants, and the atmosphere.

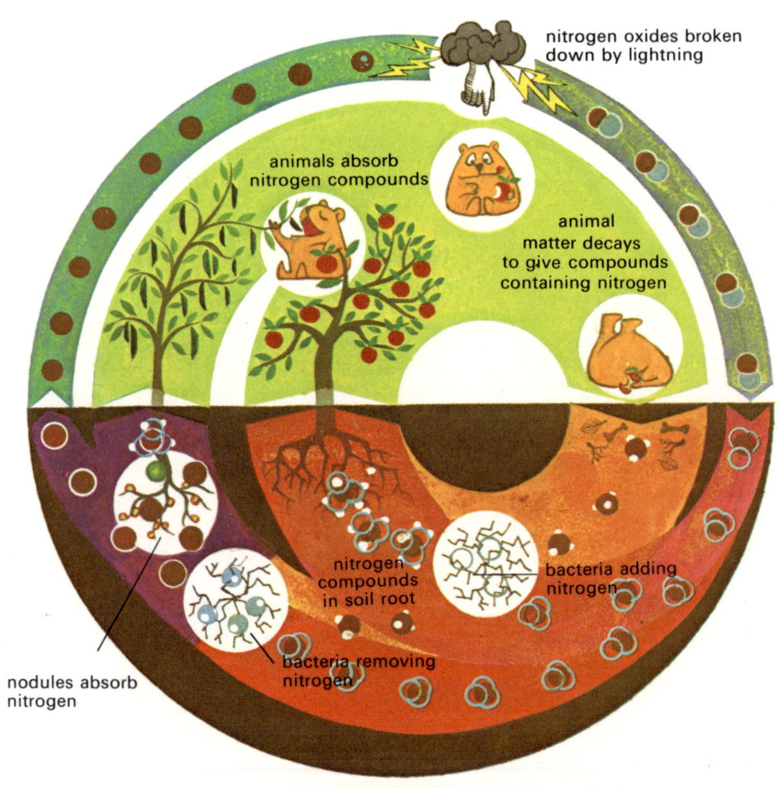

about 78 per cent of the total volume occupied by the gas nitrogen. Nitrogen is fairly inert which means that it does not easily combine with other materials in chemical reaction but it is very important when it does combine because of the variety of substances formed that are so essential to life. These substances are the *proteins*. There is about 20 per cent of oxygen in the air which is the gas so important for us to breathe. There is also about one per cent of the gas carbon dioxide which plants need. The other one per cent is taken up by a number of rarer gases.

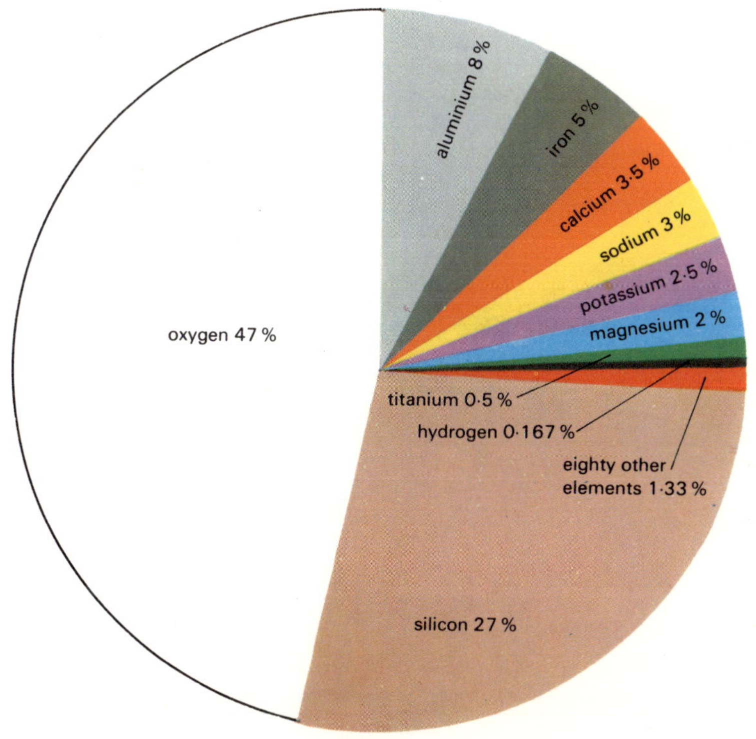

Left:
This diagram shows the main elements found in the Earth's crust.

Almost 99 per cent of the Earth's surface rocks are made up of eight basic substances or *elements*. These are oxygen, silicon which is the important component of glass, aluminium found in clays, iron which you are very familiar with, calcium which you need to form your bones, sodium found in common salt, and potassium and magnesium. Apart from oxygen and silicon all of these are metals. We shall look at the depths of the Earth in the next question.

What would you see on a journey to the centre of the Earth?

In the previous question we looked at what the surface materials of the Earth were made of. While we know a good deal about the composition and structure of the interior of the planet it is more difficult to understand how we can gain such knowledge.

We have seen when we talked about earthquakes and what happens to the waves from them, that these waves may be bent or even blocked altogether by the different materials making up the depths. Another very important piece of evidence in our efforts to build up a thorough understanding of our planet is the density of the Earth.

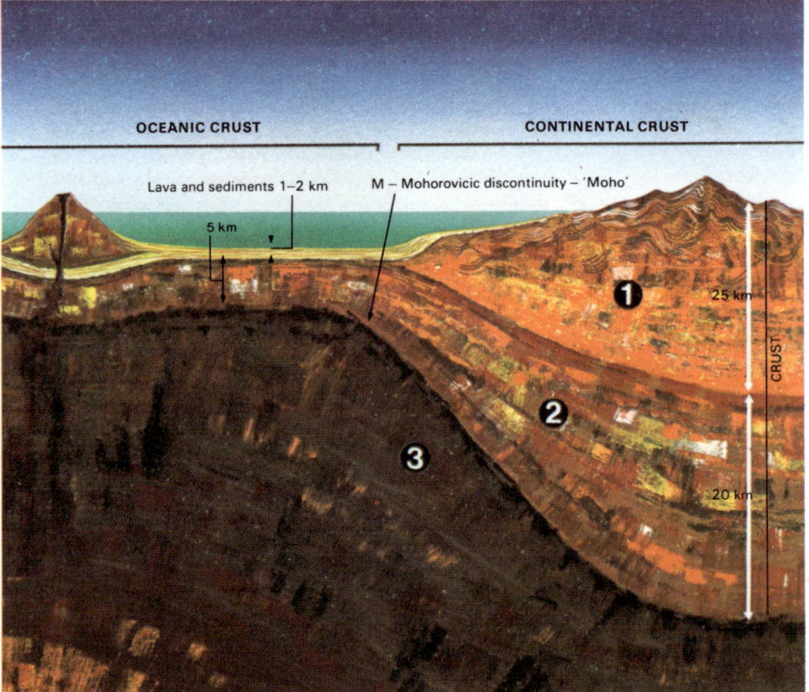

Right:
The crust of the Earth is made of:
(1) a layer composed mainly of granite, and
(2) a layer composed mainly of basalt. Beneath this, there is
(3) the much denser mantle.

Density simply means the weight of an object divided by its volume. The density of the surface rocks can easily be measured at an average of about 2.7 grams per cubic centimetre, but the density of the Earth as a whole seems to be about 5.5 g/cm^3. This clearly means that the materials which make up the interior of the Earth are very much more dense than the rocks on the surface.

There are two other clues. Firstly, the lavas which are spewed out from volcanoes and other types of rocks which arise from greater depths are an indication of what is going on at quite shallow depths. Finally, meteorites are thought

Right:
A look into the centre of the Earth.

to be composed of materials which are representative of planetary composition. But it has been the coming of earthquake studies (seismology) that has given us such a clear picture of the Earth.

The Earth is generally divided into three main zones. The surface zone is the *crust* which has been found to be about 12 kilometres thick under the ocean floor and about 35 kilometres thick under the continents. Below this there is a zone which we usually call the *mantle* and may also be called the zone of heavy rock.

The heavy rock is called peridotite and is composed of minerals containing mostly elements such as silicon, magnesium, and iron. This zone continues to a depth of about 2900 kilometres and has a density of 3.4 g/cm^3. In fact the crust and topmost layer of the mantle have often been given the names sial and sima. The sial is composed of light rocks such as granite and its name comes from the chemical symbols for silica and alumina (silicon and aluminium oxides) which make up these rocks. The sima is composed of dark heavy rocks such as basalt containing silica, iron oxides, and magnesia. The centre zone of the Earth is called the *core* which has metallic properties probably composed of metals such as iron and nickel under very great pressures.

Why does the Earth look so young?

As we have seen, the Earth has kept many secrets about herself. Geologists have been able to learn some of these secrets by looking carefully at the way in which the processes within and at the surface of the planet are constantly in motion. One of these secrets has been called the Earth's secret of eternal youth. This really means that while, for example, a person is born, grows up, grows old, and eventually dies, the Earth seems to look as young as it always has, at least, during man's short stay. This does not mean that the Earth always stays exactly the same. On the contrary, it is changing all the time.

Have you ever looked into a kaleidoscope? If you have you will know that the ever-changing patterns as you turn it are made up of different combinations of the same pieces of coloured glass. In the same way, the basic building blocks of the Earth are the same now as they ever were and the only thing that is being added is energy in the form of the Sun's rays. This is the fuel of the Earth's 'boiler'.

But if the materials stay the same, why does the Earth not grow old and die? In fact, the Earth is ageing, but this is happening so slowly as far as we are concerned that it is not easily noticeable. Of course, geologists can look back into the past and see that over a very long time changes have taken place to the planet as a whole and especially in any one area of the globe. But these changes take place in the way which scientists call *geological cycles*.

There are lots of different geological cycles. For example, the Sun's warmth heats up the sea, driving some of it up into

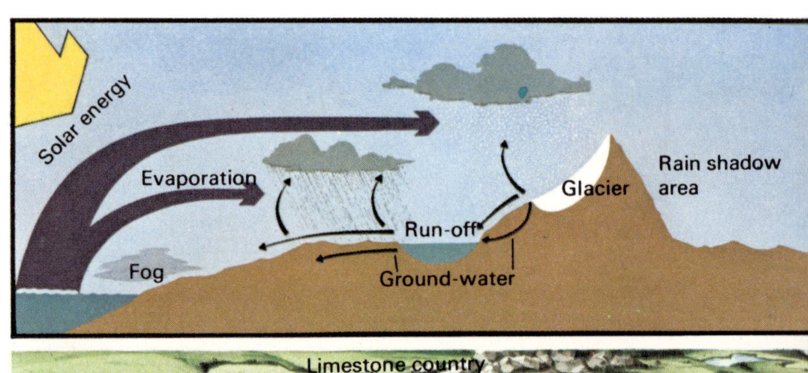

Right:
This diagram shows the water cycle.

Right:
The famous Grand Canyon in Arizona, North America is almost 2000 metres deep in some places and shows what effects a river can have on the landscape.

the sky as water vapour. (This is called *evaporation.*) This water comes together into the clouds we see, and then falls as rain, some on the land and some straight back into the sea. The rain that falls on land forms the streams and rivers and again returns to the sea, so completing the 'cycle'.

Other cycles wear away mountains and build new ones elsewhere, or take animal wastes into the soil to fertilize plants which feed the animals, and so on. We need these cycles which keep the Earth looking so young because through them many of the materials we use in our industries are brought to the surface, reborn, or even made by them. Of course, some of these cycles take millions of years to complete and we may consume their products in our factories in hundreds or sometimes even tens of years.

Below:
A landscape sculptured by some of the surface processes such as the effects of water, wind and ice.

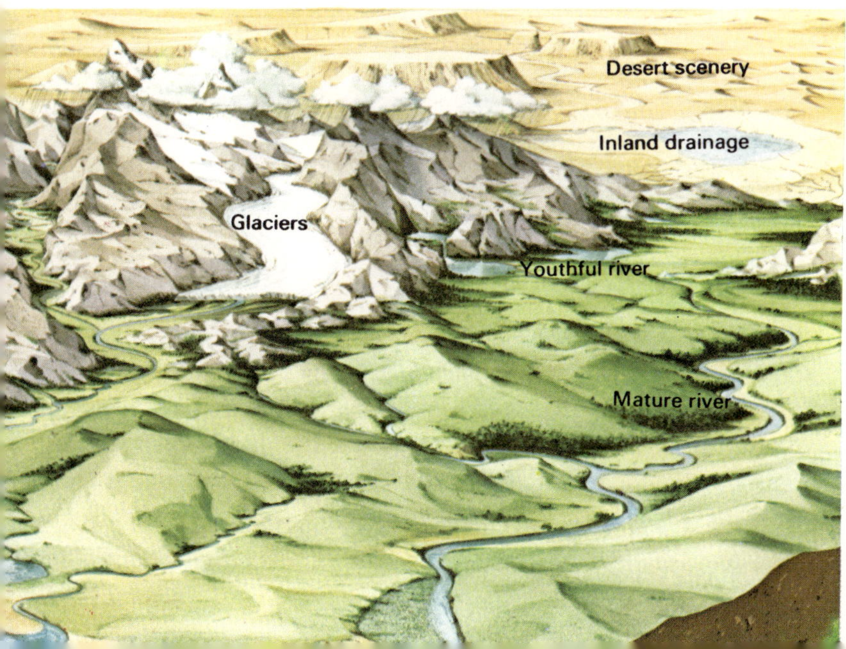

Right:
As you know, the movement of water between the land, the atmosphere, and bodies of water is the very important water cycle.

What makes the rain fall?

You will remember from the last question that rain plays an important part in helping the Earth to keep her youthful face. It is only one stage in one of the cycles mentioned, and is called the *hydrologic* cycle. But how does water rise into the sky to form the many different clouds you know so well and why does it fall again as rain?

The water, or water vapour as it is more scientifically called, gets into the air by *evaporation*. Plants also lose water through their leaves and stems, and this follows the same path into the sky.

Before rain can fall the water vapour must first cool so that it can *condense*. This means that it turns into water droplets. When you have a bath some of the steam that rises from the hot water forms a thin film of water on any of the cold surfaces in the bathroom, such as the mirror or the window. It is a little more complicated than just cooling, however. Water vapour needs to have something to form on so that it can condense into droplets. In the air this is usually the tiny particles of dust which are ever present. This means, of course, that our industries can affect the rainfall in an area because of the smoke and other substances which usually accompany them.

The clouds which develop from these water droplets are of many different kinds, and not all of them will give rise to rain. You may have noticed some of the kinds of clouds. They have interesting names like stratus or cumulonimbus.

If you have spent a holiday in an area where there are high hills or mountains, it is likely that it will have been cloudy or have rained a lot. This is because the air which flows across the hills is forced rapidly upwards by them, causing it to cool as it expands. And if this air is very moist as it is, for example, when it reaches the Lake District of England, the water quickly condenses so that this area will usually have a very high rainfall.

You have seen how the clouds occur, but this does not explain why this water should suddenly fall as rain, or even snow or hail. The main difference between a cloud droplet and a raindrop seems to be size – a cloud droplet is very tiny and floats in the air and a raindrop is larger and heavy enough to fall as it is pulled by gravity. It is not easy to see why the droplets should grow to raindrop size but one reason may be that there are always a few larger droplets present which may fall and bump into other droplets to grow still larger, and so on.

Below:
Raindrops may grow by bumping into one another and may eventually fall to the ground.

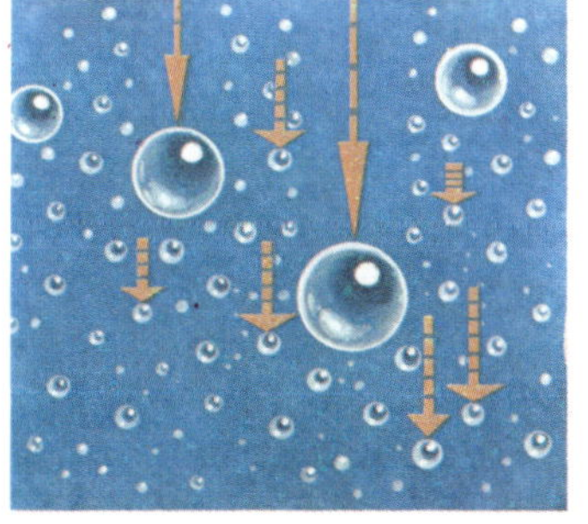

Right:
When water falls as snow, ten centimetres of snow would be equal to one centimetre of rain.

Why do the winds blow?

Perhaps you have begun to wonder not only why do the winds blow, but also why in a book which is trying mainly to answer geological questions that we have bothered to look at the rain and wind at all. We have already said that in geologic cycles whole mountains can be worn away. The rain and the winds are two of the tools which the Earth uses to do this, and it seems valuable to look at how they work. After all, it is important to understand that the many sciences such as geology and meteorology (the study of the Earth's weather) are really different parts of the same study, that of the Earth.

Wind, which you feel blowing cold on your face on a spring morning or scattering the leaves from the trees in autumn is simply the movement of air. But why should the air move at all? What is driving it? For the answer we must begin by looking again at the Sun.

The Sun's warmth is the first cause of the air's motion. The Sun's rays may warm different parts of the Earth and its surrounding envelope of air by different amounts. The amount of warmth depends upon how much dust or rain is present, the colour of the Earth's surface at any point, how much plant life there is, how densely populated it is, and so on. Thus a 'packet' of air can be warmed more than the air around it. (Hold objects of different materials the same distance away from a source of heat and you will probably find that one feels hotter than the other after a short time.)

When a packet of air becomes warm it becomes lighter and tends to rise. You may have seen films of balloons filled with

Above:
When the wind blows hard in the same general direction for some years, the shape of the trees may be affected.

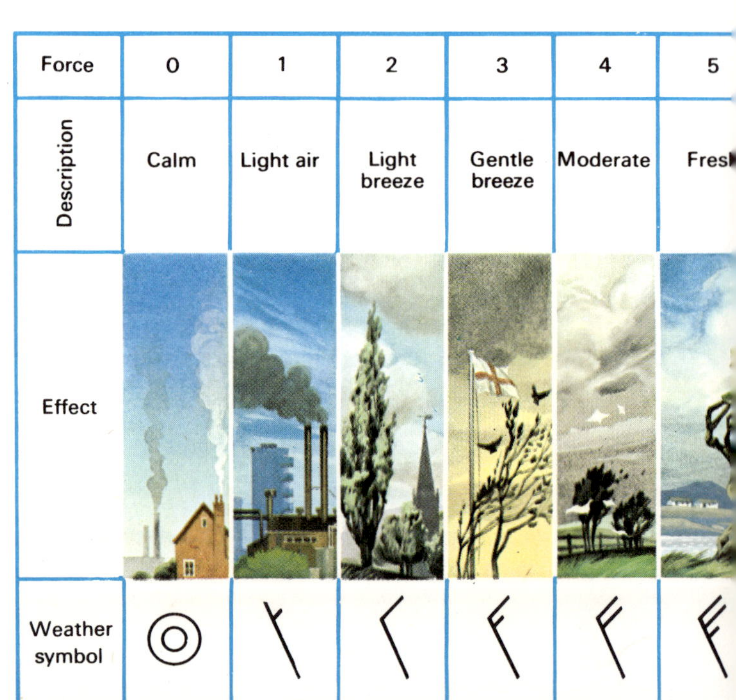

Force	0	1	2	3	4	5
Description	Calm	Light air	Light breeze	Gentle breeze	Moderate	Fres
Effect						
Weather symbol						

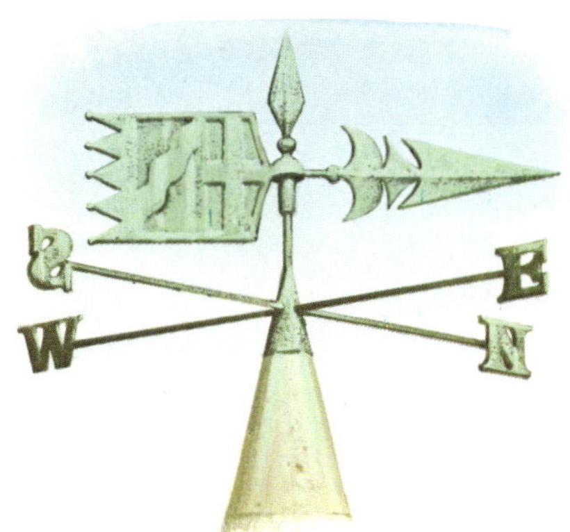

Above:
A wind vane shows the wind direction.

Below:
Frosts may be more severe during high winds.

hot air rising to great heights. As it rises the air's temperature tends to drop and this drop in temperature may cause the packet of air to descend too. So the air is in motion.

You also know that it is colder at the north and south poles and hotter at the equator. Air tends to rise at the equator and fall at the poles so that at ground level, winds would tend to blow from north to south in the northern hemisphere and from south to north in the southern hemisphere. Unfortunately, as you realize from experience, it is not as simple as this because the Earth itself is in motion, dragging the air with it. And of course, the surface of the Earth is not smooth so that the winds will be diverted by mountains and valleys. Finally, the Earth is not made of the same material all over as we have already mentioned. All these things lead to the very complicated way in which the winds blow around the planet.

Below:
The Beaufort scale of wind strengths.

6	7	8	9	10	11	12
Strong	Moderate gale	Gale	Strong gale	Whole gale	Violent storm	Hurricane

Below:
Winds may be very effective agents of erosion in deserts.

Why is it hard to make tea on a mountain?

You know that there is more to making a really good cup of tea than just pouring hot water on to a few tea leaves and then pouring it out. The next time your mother makes tea, watch. You will find that she puts the kettle on to heat, and before the water boils she pours a little into the teapot to warm it. She will then dry the pot, carefully put in the correct amount of tea, and when the water is boiling hard she will take the pot to the kettle before pouring in the water. Why does she go to all this trouble for a cup of tea? The answer is that you simply cannot make tea if the water is not at a temperature of 100°C, that is, under normal conditions when the water is boiling. But what is the connection with making tea on mountains?

To understand this, you must know what is meant by boiling. All liquids release a certain amount of vapour. This vapour exerts a pressure; that is, it pushes on its surroundings. As a liquid is heated, the pressure increases until eventually it is the same as the pressure of the air – this is the boiling point.

Below:
The atmosphere pressing on the surface of the mercury is enough to force the mercury to rise about 760 millimetres or to support a column of mercury 760 millimetres long.

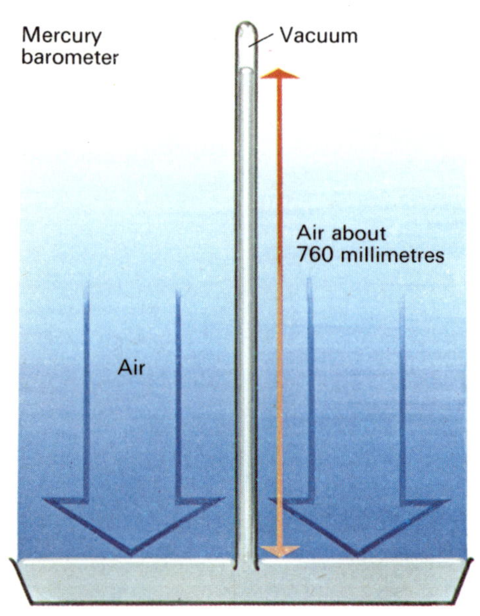

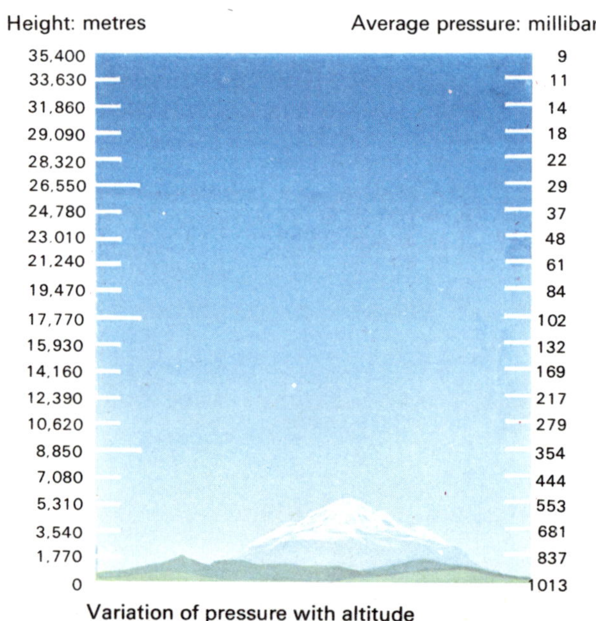

Variation of pressure with altitude

Above right:
This diagram shows how atmospheric pressure varies with altitude.

What is air pressure? We do not feel the air pressing down on us. But in fact, every square centimetre of the Earth's surface at sea-level feels an air pressure of about one kilogram. This is really the weight of a column of air measuring one centimetre by one centimetre by the height of the column.

Right:
A household barometer indicates the weather conditions.

And that is quite a lot of air!

What happens when you climb a mountain? Obviously, the column of air becomes shorter and the pressure of the air is reduced. For example, if you were to climb to a height of 6000 metres, the air pressure is about half the air pressure that you would feel at sea-level. If you try to make tea at this height you will find that because the air pressure is lower, the pressure of the water will be lower when the water boils and it will not be so hot. You cannot make the water any hotter because it simply turns into steam. This all leads to a rather poor cup of tea!

If you have watched the weather forecast on television you will have seen that the forecaster's charts have lots of lines on them. The air pressure along any one line or *isobar* as it is called is exactly the same, but different lines indicate different pressures.

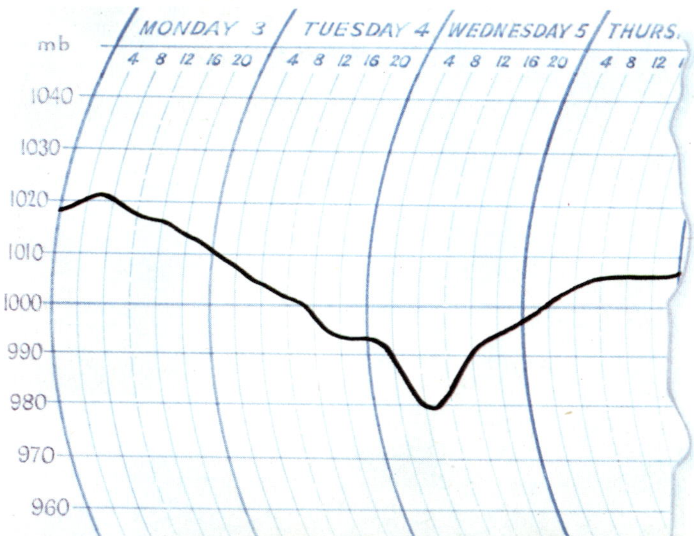

Right:
Some barometers record the variations in pressure from day to day on a chart like the one shown here.

Can a rock rot?

Above:
A visit to a cemetery will prove to you that rocks can rot.

The answer is yes! If you have ever looked at a china clay pit and seen the soft, white, sticky china clay it is hard to imagine that it was once hard rock. For example, the china clay found in South Devon, England is the result of the rotting of the granite underlying the moors of Dartmoor and Bodmin.

But what happens to make the granite rot into clay? Rocks rot by the process known as *weathering*, and all rocks suffer from its effects, whether they are found in the tropics or the frozen wastes of the Antarctic. Weathering is the sum total of all the effects on the surface materials caused by the processes of heating and cooling, freezing and thawing, and so on.

There are two main types of weathering given special names by geologists:

1. **Physical Weathering** These are the processes by which rocks are broken down by actions such as change of temperature on the materials themselves, and also the effects of water freezing and thawing in contact with the rocks.

2. **Chemical Weathering** This is the term used to describe what happens when the chemicals that make up the rocks change as a result of the action of water and air.

It is also worth noting that plant growth can have a

Below:
Trees and other plants can help to prise rocks apart.

Below right:
The freezing and thawing of water in joints can cause rocks to shatter. Heating and cooling can lead to 'onion skin' weathering.

Right:
China clay is the result of the natural weathering of granite.

considerable effect on rock materials, as can *bacteria*– tiny organisms which are not like any other plant or animal. In fact, without these bacteria which release essential nourishment from the rocks there could be no more advanced life forms, including man himself.

Physical Weathering
In areas of the world where it is damp enough to have rain and becomes cold enough for the water to freeze, rocks can be prised apart by the action of ice as it freezes and expands. In hot, dry deserts rocks may become very hot during the day only to cool very quickly as soon as the sun goes down, so that they expand and shrink suddenly. This can cause the rocks to split with explosive force.

Chemical Weathering
When rain falls it may carry gases from the air such as *oxygen* and *carbon dioxide*. The carbon dioxide makes the rain slightly acid; acid enough to be able to dissolve rocks like limestone. The Karst district of Yugoslavia, for example, where the rocks are worn into blocks and channels and swallow holes and caves, shows the effects of this very well.

You have seen, then, that, like a piece of wood or an apple, a rock can become rotten when its materials may fall victim to rain and wind.

When is a rock a soil?

Most of us think of soil as the black or brown substance in which we grow our vegetables in the back garden; the substance which gets wet and sticky when it rains and makes our shoes dirty. But what is it that gives soil the special property which allows us to grow many of the plants that we and other animals need?

We have already seen from the previous question that rocks can break down in various ways. When the products of this breakdown are mixed with a material called *humus* the result is known as soil. If you were to look at a quarry you would probably find that on top of the solid rock there is a band of broken and partly rotted rock, and then as you move upwards there are different layers of soil. What you would actually see depends upon the type of rock, the climate and the way in which the water in the soil moves, and the type of vegetation growing there. Of course, plants also reflect the climate and the type of soil, so that you can see each depends upon the other in a very complicated relationship.

Right:
Bacteria can extract nutrients from rocks. Mosses and lichens would then be able to grow. Grasses could then grow in cracks where tiny amounts of soil have formed.

What is humus? Humus is a jelly-like acid material which results from the decay of the remains of plants and animals. You may ask where the first humus came from if humus comes from plants and animals. Animals need plants which grow in the soil. Soil needs humus. As you read in the last question bacteria can break down rock, and the substances thus

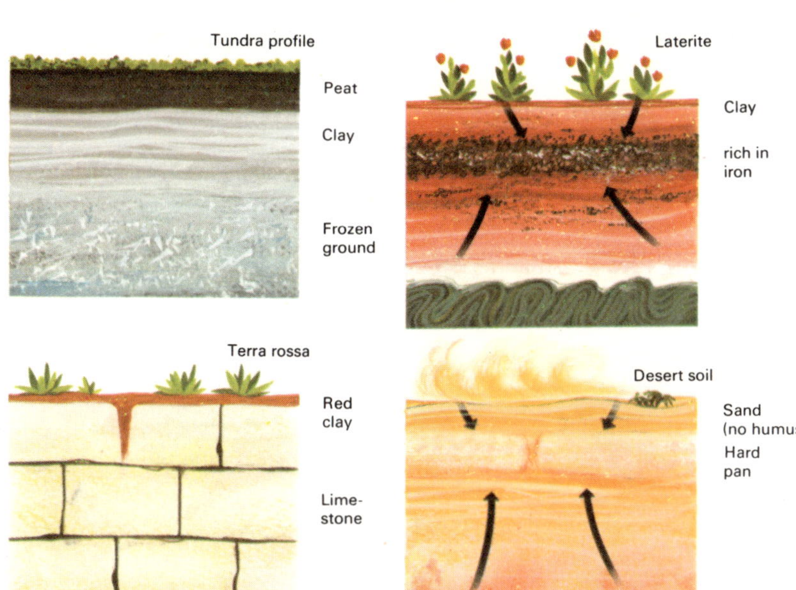

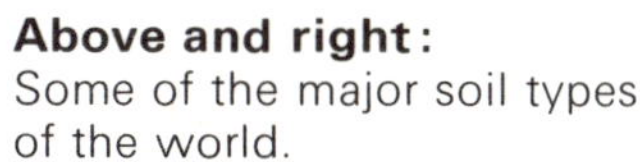

Above and right:
Some of the major soil types of the world.

Above:
A soil section.

released provide nourishment for the first specialized plants such as lichens and mosses. When these plants decayed their remains mixed with the rock particles to form the first soils.

Scientists who study the soil are called pedologists. This word is derived from the Greek, *pedion*, meaning the ground, and *logos* meaning discourse or speech. In an effort to understand the complicated world of the soil, pedologists use *soil profiles* which are simply vertical sections through old, well-established soils showing the various layers. These layers are formed as the soil ages by the action of rainwater moving material downwards (*leaching*), the concentration of humus by plant growth, and so on.

As climate influences soils, pedologists recognize that there are a number of main types of soil that can be seen in the various climatic regions around the world. *Podzols* and *brown earths*, for example, are found in Britain and Europe where the climate is temperate.

How are mountains worn away?

When we talked about geological cycles we said that mountains can be worn away. You have already learned that hard rocks can become soft and rotten by weathering and may even dissolve completely. Eventually, all the materials which result from this weathering process may be transported from their place of formation by the Sun-driven carriers such as rain, wind, or sea.

The Earth's gravity can also play an important part in removing these weathered materials, particularly on hills and mountains. On hills, helped by the action of the rain, or even by plants and animals, the soil can slowly *creep* downwards. And of course, as the soil is removed, more rock is exposed to the elements so that weathering can once more take its toll of the bare surface.

In dry deserts where there is not enough moisture to hold the weathered material together, the wind may pick up particles of sand and hurl them against any rocks which may be still standing and slowly wear them away thus providing even more ammunition for the sand-blasting wind.

Water, in its various forms, is probably the most important of the Earth's many transport systems. We have seen that rain can accelerate the process of soil creep, even though occasionally a particularly hard rock can protect what lies beneath it from further decay so that earth pillars result. But much of the rock debris, as it is sometimes called, finds its way into the rivers and streams where it may travel great distances. It is carried by rolling along the river bed,

Below left:
At first water runs down the slope in sheets; it then forms a channel and deep gulleys may develop.

Below right:
Severe rain erosion can cause the land to look like this. A landscape like this is known as *badlands*.

Bottom left:
Ripples may be formed on the river bed and particles may be carried in suspension or may be rolled and dragged along the river bottom.

Bottom right:
A dried river bed may expose pot holes.

Right:
The effect of soil creep.

Far right and below:
Landslips can cause severe erosion particularly in coastal areas where cliffs are undercut by the sea.

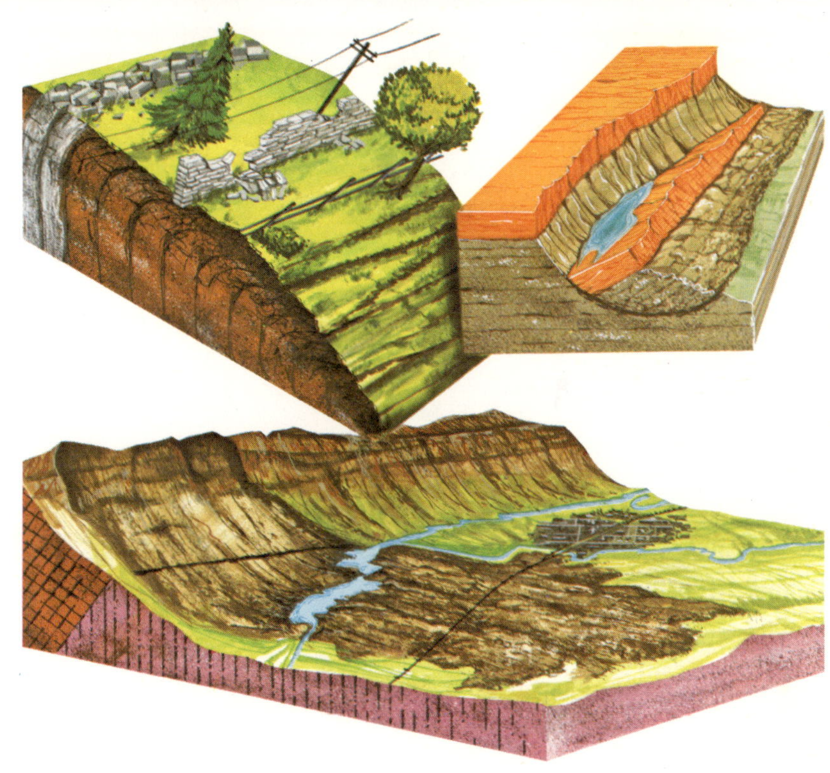

Below left:
A hard rock may protect the underlying rocks from erosion resulting in earth pillars.

Below:
These dramatic screes are to be found at Wastwater in the English Lake District.

Inset:
A raindrop can have a considerable impact on wet soil.

Below right:
Erosion of these folded rocks have caused older rocks to be exposed in the core of this anticline (*see* page 77).

by bouncing, or by complete suspension in the water itself. The dissolved materials may move also in the form of solutions.

Particles of rock wear away by bumping into one another and hitting the sides and bed of the river. They are rounded and polished as they are carried along and may further widen and deepen the river valley itself. As the river deepens its valley, gravity plays an increasingly important part in carrying material down the valley sides into the river. Ice, too, can trap rock fragments within it and carry them down valley as we shall see later. Most of the material eventually finds its way into the sea where the tides and currents may sweep it many kilometres from the original mountain or hill.

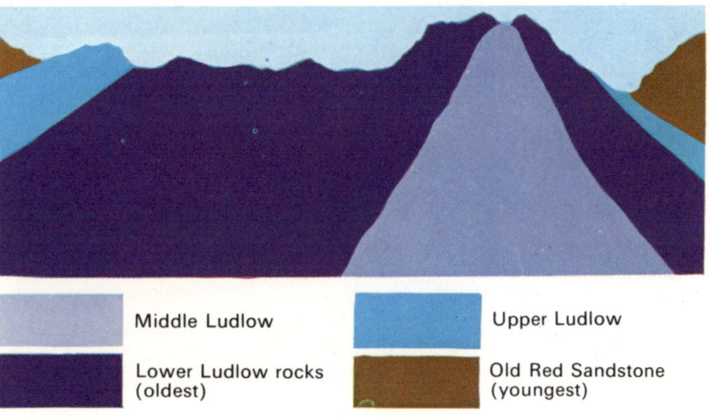

Where does all the rain go?

You have read from some of the previous questions why rain falls from the skies, and the importance of rain in wearing away rocks and transporting the debris far from its place of origin.

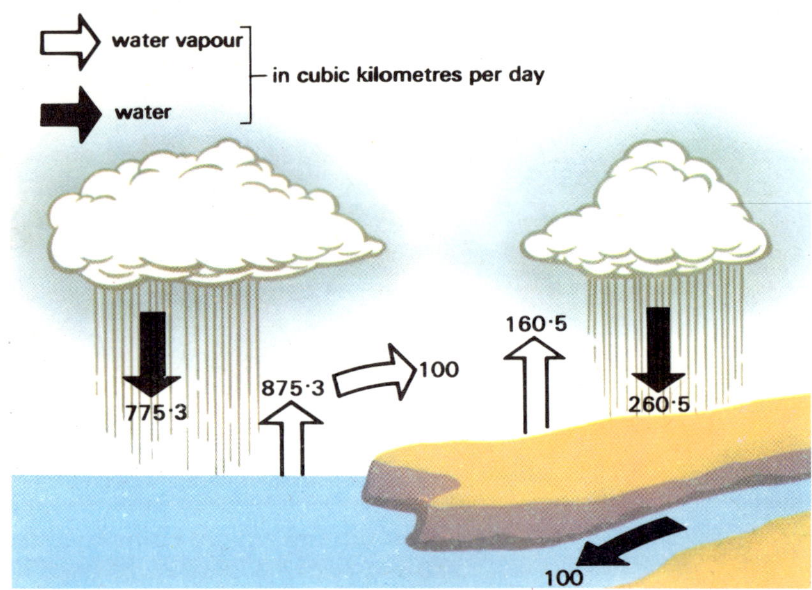

Right:
This picture shows the quantities involved in the water cycle.

Two-thirds of the surface of the globe is covered by water. This water is found in the world's oceans and much of the fresh water which is essential for all life has evaporated from them. But what happens to all the rainwater that falls on land? Some is evaporated straight back again, some soaks into the ground, some is taken up by the plants of the Earth. All the water in the Earth's rivers and streams, however, comes from *meteoric* water, that is water that falls from the atmosphere as rain. Of course, we use a great deal of the rain which falls on to the land surface, not only for washing and drinking but also in our industries and as a means of transport and disposal of our waste products. Here is another case where man's demands on the resources of the planet are beginning to show signs of wear, as you would soon notice

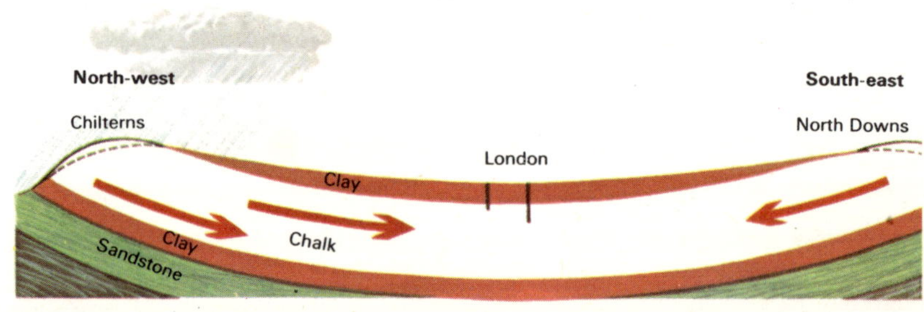

Right:
Some of London's water is drawn from wells drilled into the chalk aquifer. The rain originally fell on the North Downs and Chilterns some kilometres away.

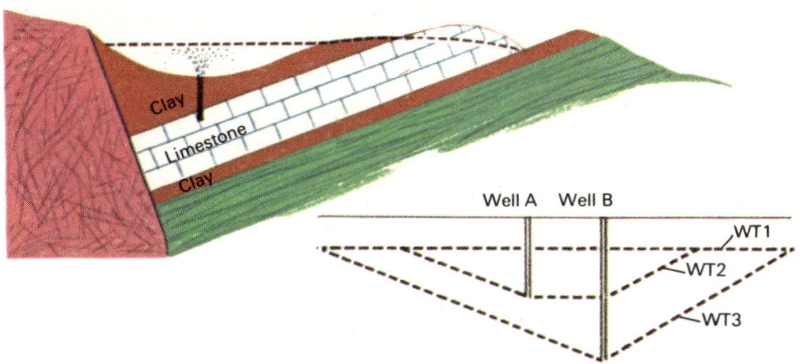

Right:
An *artesian well* is produced when the top of a drilled well is below the level of the intake to the water-table.

if you were to look at the dead, stinking waters of the Hudson river in the United States of America where fish can no longer live.

The rain which eventually forms rivers falls on to hills where it first runs downwards as a sheet of water. Soon the water cuts out channels which then come together to form larger streams and rivers. The way in which a river and its valley develop, depends upon the amount, frequency, and violence of the rain, the type of rocks and the slope upon which it falls. It depends also on the plant cover.

When the river flows down hill on a land surface which has been newly uplifted by earth movement, it is called a *consequent* stream, and the smaller streams which run down its valley sides to join it are called *tributaries* or *subsequent* streams. In this way a river system is formed.

We mentioned that some of the rain soaks into the ground – it becomes *groundwater*. Most rocks, such as sandstone, have many tiny holes between the grains which allow water to pass through them, and they may also be cracked which aids the flow. Other rocks, like clay, (clay is certainly a rock) do not allow the passage of water, and act as a barrier, so that water may build up against it forming an *aquifer*. The top surface of this water-bearing layer of rocks (resembling a wet sponge) is called the *water-table*. We drill our wells providing us with much of our water into this layer, but of course, if we take out more water than is flowing into the aquifer, the well will eventually run dry.

How can a river grow old?

We have explained how some of the rain which falls on to the land surface finds its way into the rivers and how a river system is born, but now we have a chance to look at another one of these familiar geological cycles. This time it is the cycle of *erosion*. This really means the way in which a river

Right:
This illustration shows how a river and its surrounding scenery evolves:
(1) Davis's youthful stage;
(2) maturity;
(3) old age. The dotted lines on the waterfall show the original position of the resistant rock indicating how the fall has been cut back. It is not certain why meanders develop, but they may become so extreme that the two parts of the meander neck may join, leaving the cut-off channel as an ox-bow lake.

system works upon and moulds the land forms. It has been suggested by an eminent scientist called W. M. Davis that there is a sequence of stages. He suggested that a river valley system can pass through three major stages provided that there is no interruption by further uplift of the land. He called these stages *youth*, *maturity*, and *old age*.

In practice, it is now thought not to be quite as simple as this, but you would still be able to recognize these stages if you saw them in the countryside. Davis suggested that in the youthful stage when the land has only just been uplifted the water rushes down the steep slopes carving deep, narrow valleys with many waterfalls.

Eventually the rate at which these valleys are carved slows down and the river becomes wider, particularly where the rocks are less resistant to this wearing away. At this stage the action of the main rivers and their tributaries have begun to wear down and flatten the original surface. This is the mature stage.

At old age the landscape has been worn to that of a flat or very gently sloped region called a *peneplain*. The main rivers wind their way slowly towards the sea in a series of loops called *meanders*. Those hills which do survive are called *monadnocks*. This time of old age, or the *senile* stage as it is sometimes called, may take a very long time to develop fully and in addition to the meanders you may be able to see that the river channel may eventually become raised above the level of the surrounding plain. The embankments holding back the waters are made up of fine silty or sandy material called *alluvium* deposited there during times of flooding. They are called *levees*.

Sometimes, the meanders may curve around so far that the river finds it easier not to flow around the bend and the neck of the bend is cut off, leaving the remaining bend as an *ox-bow lake*.

Finally, it may happen that the sea-level may fall or the whole or part of the peneplain may be uplifted again and the river may be *rejuvenated*. That is, new rapids may be born, and the meander channels may be cut much more deeply by the fast flowing waters which then flow down to the sea.

Right:
This illustration shows some of the land features associated with river development.

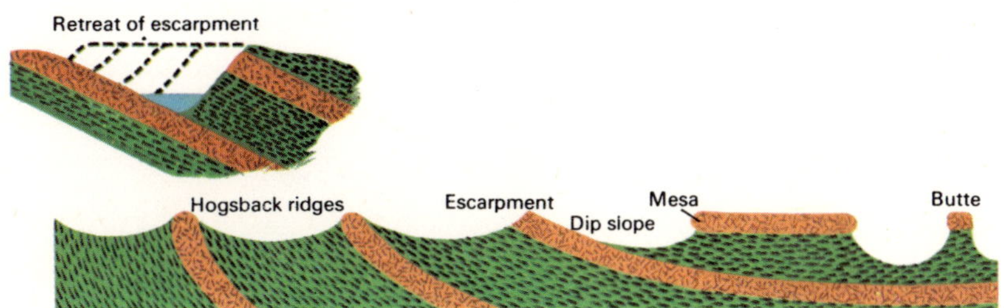

What are glaciers and icebergs?

Top:
The extent of the ice in the north polar region today.

Above:
Icebergs in the north polar seas today.

You have probably been told about the huge masses of ice that exist at the north and south poles today. You may also have heard that thousands of years ago great tongues of ice stretched southward from the north as far as Britain and northern Europe, and that millions of years before many of the countries of the southern hemisphere were similarly affected by an *Ice Age* as these periods of extremely harsh conditions are called. But how do these tongues of ice occur and what effects can they have on the face of the Earth?

It has already been mentioned that some of the water falling from the sky may do so as snow rather than rain. This is particularly likely to happen at the tops of mountains and in the regions closer to the north and south poles, where the temperatures are generally lower. Of course, we quite often wake up on a cold winter's morning to find everything outside shrouded by a glistening, soft, white mantle, but our snow soon melts again. In some parts of the world, however, such as in Greenland or the Himalayas, not all of the snow which falls in the winter is melted during the short, cool summer months and the snow tends to build up into *snowfields*. This is the first condition necessary for the formation of the frozen sculptors of our planet, the *glaciers*. As the snow builds up in the snowfield, it becomes compacted into bluish ice rather than the familiar, white snowflakes.

Right:
A glacier flowing down from the snowfield.

Far right:
During the last Ice Age the ice, shown in white, extended much farther than it does today.

Inset:
This diagram shows how the centre of a glacier flows faster than it does at its margins.

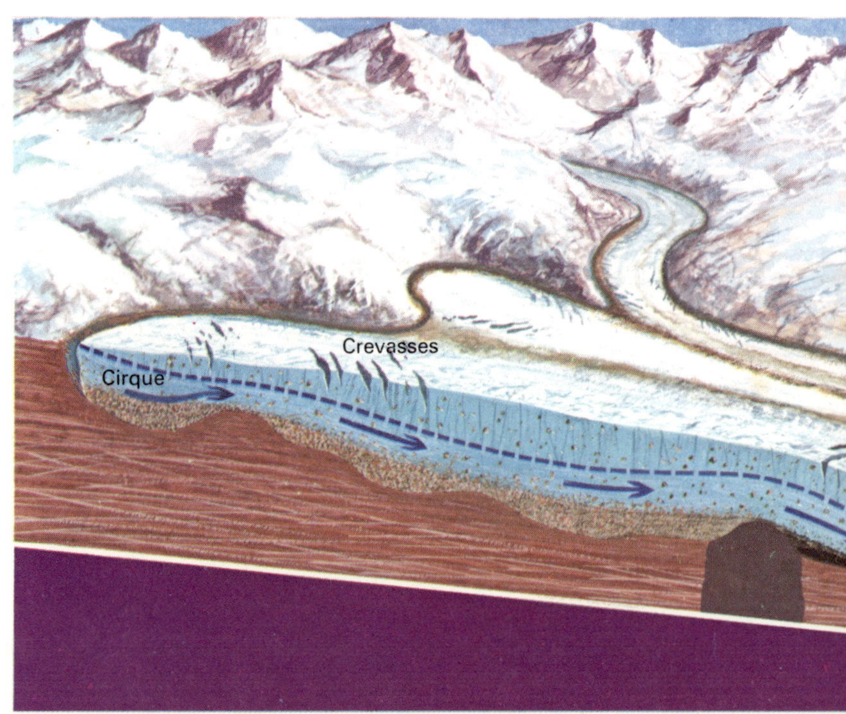

The glaciers often flow down-hill from the snowfields where they are formed and only come to rest when the rate of melting of the ice-fronts, or snouts, as they are often called, is greater than the amount of ice which can be provided by the snowfield.

When the snout of a glacier meets the sea as often happens in Alaska, for example, great chunks of ice may break from the parent glacier by a process called *calving* to send an *iceberg* out into the icy seas. If you have seen photographs of these great blocks of ice you will know what an impressive sight they are, but it is as well to remember that the mountain of ice which you see rising grandly out of the water is only the tip of the iceberg, and that nine-tenths of it is under water. It is not surprising then that it was an iceberg that was responsible for the tragic sinking of the great steam ship which was thought to be unsinkable, the *Titanic*. She foundered and sank very quickly on her maiden voyage with the loss of over 1000 lives after a collision with an iceberg.

Top and above:
This illustration shows the extent of the ice in the south polar region today. Note that penguins only live in the south polar region.

Left:
Parts of Europe may have looked like this during the last Ice Age.

Can glaciers move mountains?

Try taking an ice-cube from the refrigerator and rubbing it against a stone from the garden. You will find that not only does the ice melt where it has been in contact with the warmth of your hand, but that it also quickly melts where you have been rubbing it, and it has no effect on the stone. You might think, then, that this is rather a silly question. Of course, you know that glaciers and icebergs are much bigger than ice-cubes, and that it is much colder in their surroundings than in your garden. But even the largest glaciers can only push about a fiftieth as hard as would be necessary to break a piece of granite.

In reality the movement and accompanying processes of a glacier in a valley are much more complicated than that of water. In the warmer regions where large amounts of ice still occur there may be a lot of water melting from the ice which might trickle into the underlying rock and on freezing again would exert enough pressure to shatter it. There are many scientists who study glaciers, however, who

Right:
Louis Agassiz was the first geologist to study the extent of glaciation in former Ice Ages. This illustration is based on a sketch he made in about 1841 which he used in connection with his early studies.

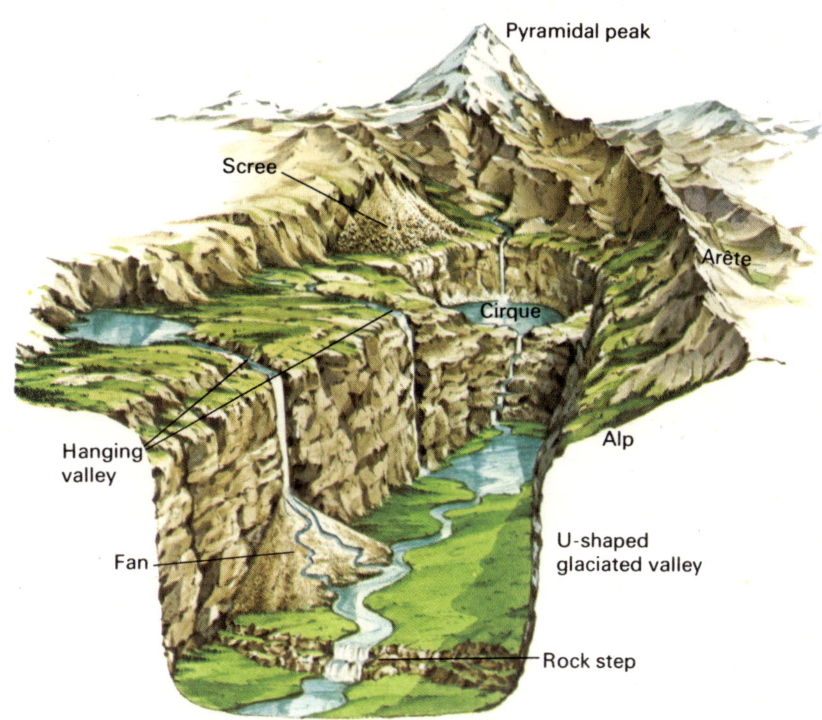

Right:
These are just some of the features which a glacier can leave behind as marks of its power of erosion.

believe that this cannot happen. In any event, these effects are only important in the top few centimetres of the rock and would not account for the removal of a significant amount of rock.

You know that if you compress a spring and then release it, it springs back into its original shape quite violently. In a similar way, all rocks have forces stored within them like a compressed spring, and by removing the top few inches of surface rock these forces may be released with such vigour, so far as the rock is concerned, that the rock may shatter still further. This allows the ice to pick up and carry away more material.

We have said that rock debris is removed by ice. This occurs because as the ice presses upon the fragments of rock, the ice tends to melt forming a cavity. The fragment lodges into the cavity and the ice eventually carries it away. These suspended stones can also aid the erosive power of the glacier in the manner of a piece of glass- or emery-paper. If you have built balsawood model aircraft, you know the way in which you smooth and round off the leading edges of the wings by rubbing them with glass-paper. In a similar way, a glacier may carry a large block of rock which might easily break off other rocks jutting from the valley wall, so that smaller fragments can begin their task of smoothing and polishing the remainder.

How do you know when ice has been at work?

As we have already explained, in certain parts of the world, snowfields are formed and glaciers extend from them. These glaciers wear away the valleys in which they are confined, transport rocks, and deposit clay and rocks when they melt. In many ways the glaciers affect the landscape. In the past, sheets of ice were sculpturing the surface of the Earth in areas which are quite mild in climate today. Areas such as the English Lake District, Snowdonia of North Wales, and the Alps are all good examples of glaciated landscapes. But what features do all these areas have in common to allow us to tell that they have been affected by ice?

An area that has been glaciated is marked out by two distinctive sets of land features:

1. Features resulting from erosion of the area by the glaciers.

2. Features which result from deposition of material which had been carried by the ice.

We shall deal with features resulting from erosion first. The ice may carry rock fragments within it and those in contact with any rock surface will scratch it as the glacier passes over the rocks. These grooved rocks are called *striated surfaces*. If some of the rocks are harder than others, they may remain raised above the valley floor as ice-moulded hummocks which are smooth on the side facing the direction

Right:
A glacier may also leave behind material which it has been carrying in characteristic forms such as those shown in the illustration.

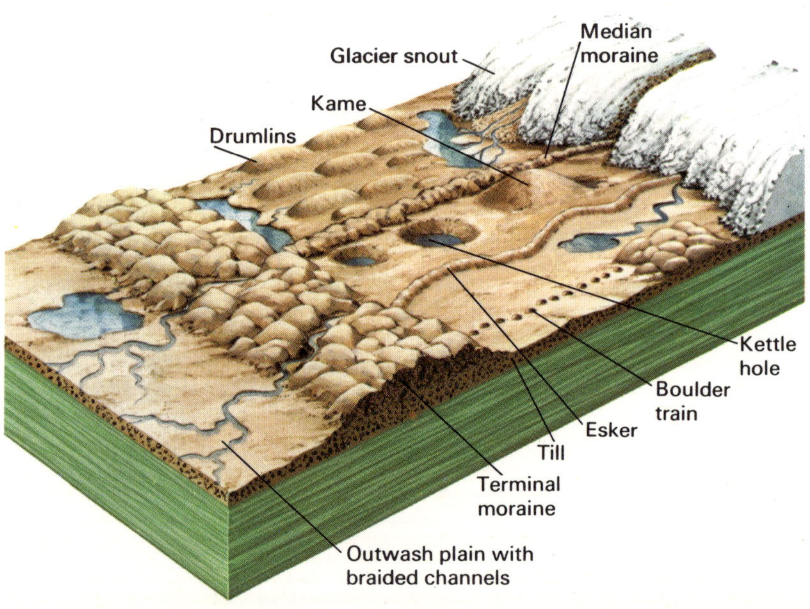

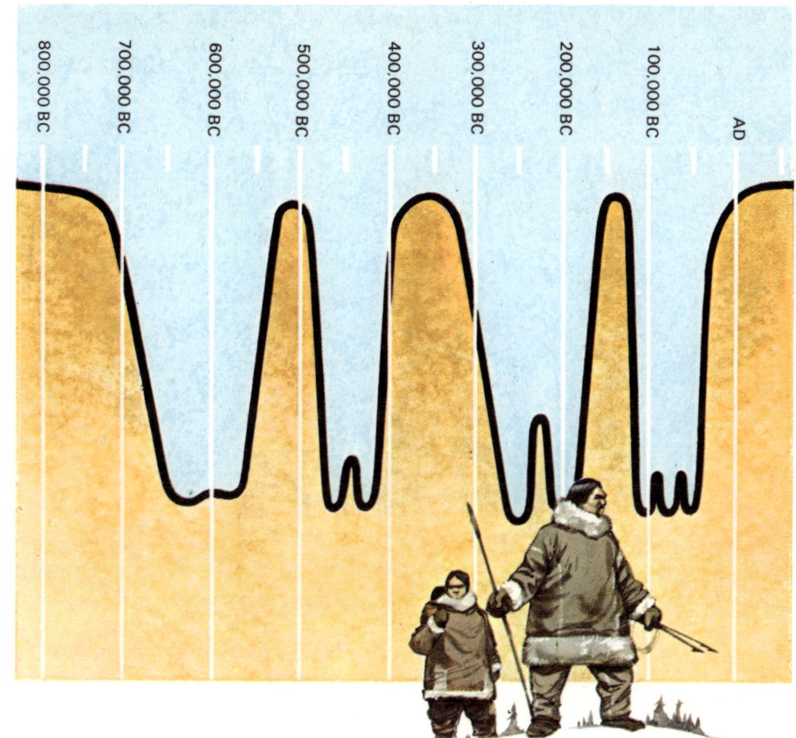

Right:
There have been four glacial phases or Ice Ages during the last 800,000 years as indicated by the troughs in this picture.

of the flow of ice and jagged on the other side. These are called *roches moutonnées*.

Snowfields formed in mountain hollows are cut back and deepened by the processes which are associated with the ice and then small glaciers move downslope. This commonly happens in existing valleys. The armchair-like hollows which are formed at the head of the valley are called *cirques, cwms,* or *corries*. When two or more of these cirques occur on different sides of a mountain, the erosion may cause them to meet, and an *arête* or a *pyramidal peak* will remain. The Matterhorn in the Alps is a very famous peak of this kind.

The glacier moving down the valley will cause the valley to become deeper and U shaped leaving the original tributaries hanging, and any spurs left by the meanderings of the river will be worn away or *truncated*.

Deposits of rock fragments and clay left behind after a glaciation may be dropped from the ice itself or from any of the melt waters accompanying it. The haphazard mixture of rock and fine clay is known as *drift* or *till*. It may be dropped at the nose of the glacier as it melts giving rise to *moraine*. Sometimes a block of one kind of rock may be carried many kilometres only to fall on a completely different type of rock. This is an *erratic*.

These and many other features give us very strong clues that ice has been at work.

How do we know that there were deserts in Europe?

When we answered the question: 'How are mountains worn away?' we explained how the effects of the wind are so important in picking up the weathered material. We did not, however, explain why the wind should be so much more important in desert regions.

When you think of a desert, we expect you think of the great deserts like the Sahara in North Africa or the deserts of Arizona in North America, or even the interior of Australia. What do these regions have in common? The first thing that springs to mind is that they are very hot, and you might imagine that all deserts are hot. In fact scientists do not take the temperature into consideration when an area is called a desert. A desert is simply an area which receives less than 250 millimetres of rain in a year. They may be hot or cold but they are all very dry. The Arctic is a desert.

In areas with a more normal rainfall, such as Britain and Europe, there is enough moisture on the land surface to prevent the wind from being able to pick up and carry away even quite small particles of rock or soil, so erosion by wind is not important. In more recent years, however, the removal of hedges from areas such as East Anglia, England, by farmers who wish to make their fields larger and easier to plough has led to some wind erosion. This is because the hedges act as wind breaks, and also help to tie the soil

Right:
A typical modern desert.

Right:
Some typical dreikanter.

Right:
These sand dunes are called *barchans.*

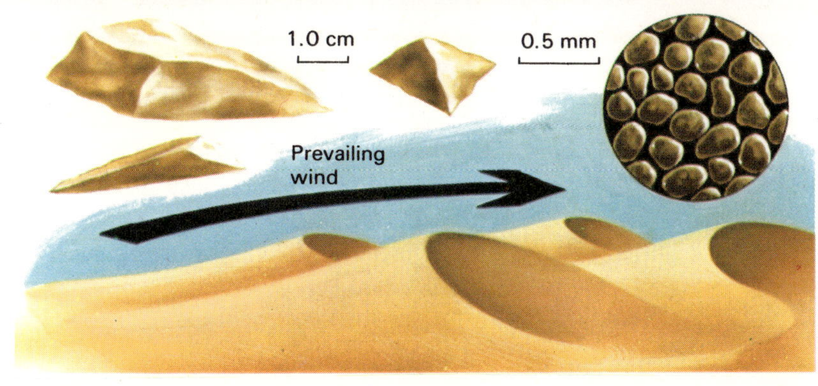

together with their roots. In Germany, farmers are being encouraged to put back the hedges which they once removed.

In the dry deserts the wind can pick up the dry, loose deposits which have resulted from weathering, and may lower the land surface. The particles picked up are thrown against each other and against any upstanding rocks. This process is called *wind abrasion*. The particles themselves soon become rounded but their surfaces become pitted like frosted glass. Isolated pebbles may become worn on the side facing the blasts of sand-bearing wind. If the wind direction changes, other sides of the pebbles will be ground down. If this happens three times, a three-faced stone is the result, and is called a *dreikante*.

What happens to the sand that is constantly on the move in the deserts? It forms *dunes*. These are hills of sand similar to the ones that you have probably seen at the seaside. The sand is built up in layers which are always on the move following the direction of the wind.

If, then, we were to look at any rock today, and found signs of any of these and other features we would know immediately that there was once a desert there. There are sandstones in Ayrshire, Scotland which show many of these characteristics.

Some of the special conditions that can be seen in modern deserts.

Right:
A fan left behind by a river which has since dried up.

Right:
A wadi or dry valley which may become a raging torrent for a short time when the rains come.

What are tides?

If you live by the sea or you have had a holiday at the seaside, or even if you have spent some time by the mouth of a river, you will know that at certain times of the day the waters rise and fall. These ups and downs of the seas are called *tides*. In fact, on average, the rise and fall occurs every twelve hours and twenty-six minutes.

Why should tides occur? You know that everything on the Earth's surface is attracted towards the Earth by a force called *gravity*. Actually all objects are attracted towards each other by this force by an amount which can be worked out by a simple rule called the *inverse square law*. Roughly, this means the greater is the distance between two objects the smaller is the attraction, but the larger the objects the greater the attraction.

As the Moon passes around the Earth it attracts the waters of the oceans on the side facing it and causes them to bulge. On the opposite side from this bulge there is another one. Because the Earth is closer to the Moon than the oceans, the Earth is attracted towards the Moon more than the waters. This means that a bulge of water is left behind. These bulges pass around the Earth with the motion of the Moon and give rise to the tides.

There are two other things which help in the formation of the tides. You know that if you cause the water in your bath to rock, it may rise and fall against the side of the bath for some time. In the same way, once the tides have begun, the waters tend to continue to rock up and down and they are given an extra push by the attraction of the Moon. The Sun also tends to attract the Earth's oceans towards itself, but

Right:
The gravitational attraction of the Sun and Moon produces the daily tides.

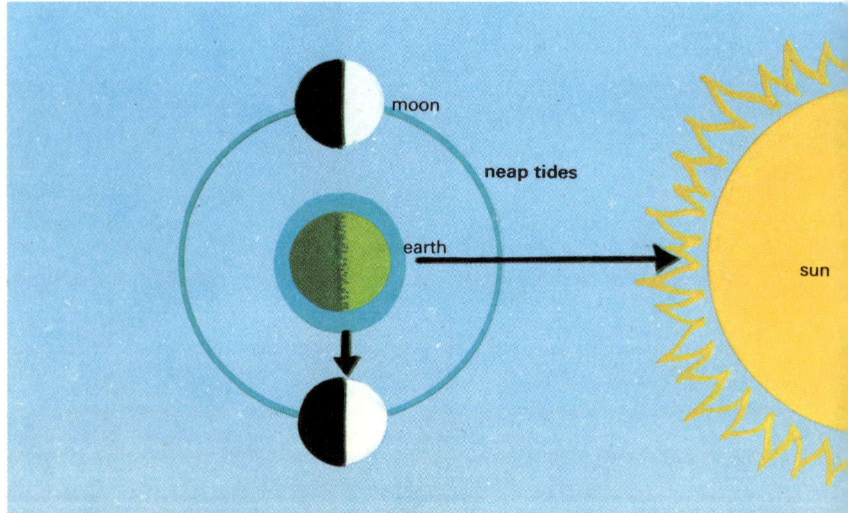

Left:
At low tide, moored boats and flotsam and jetsam may be left high and dry.

because it is so very much further away the attraction is much less important. At certain times of the year, however, the Sun, the Moon and the Earth are all in a straight line. When this happens, the attraction of the Sun is added to the attraction by the Moon, and the tides are particularly high. These are called *spring tides* and only happen when the Moon is new or full.

At other times, the Sun, the Earth, and the Moon are in positions which look like the corner of a square, with the Earth at the point. When this happens the attraction of the Sun tends to cancel out some of the pull of the Moon and the tides are much lower. These are called *neap tides*.

Why are there waves on the sea?

Only very rarely are the seas completely still. These were the times that sailors on the old sailing ships feared the most – even more than they feared the worst storms. They became becalmed, which meant that no wind was blowing to fill the great sails to carry them to their next port of call. The sea would be motionless and mirror-like and no rain would fall. If the sailors were held fast for too long their food would begin to run out, and perhaps more important fresh water would become scarce.

You may have guessed by now the connection between waves on the sea and the wind. In fact, waves are almost wholly the result of the wind blowing across the surface of the water. Perhaps you have played 'blow' water polo at home. You need a table tennis ball, two pieces of tube and a bowl full of water. Float the table tennis ball in the bowl and blow through the tubes. As you blow on to the water you will notice that the surface is stirred up into ripples – the harder you blow the bigger the ripples.

It is very similar on the sea. The wind drags the water to form waves which slowly move forward and get larger. Although the wave shape moves forward, each particle of water moves round in circles and does not change its average position. The height of a wave depends upon three factors. How hard the wind is blowing (you will realize why gale warnings are so important to sailors), how long the wind has been blowing, and the *fetch*. The word fetch means the length of the stretch of open water over which the wind is blowing.

Right:
Breaking waves can exert tremendous force and cause cliffs to be eroded away.

What happens when a wave reaches the shore?

When a wave approaches shallow water – shallow means that the depth is less than half the distance between two wave tops – it is slowed down by dragging on the sea bed. The first waves to approach the shore are slowed first and the ones behind pile up. The amount of water in each wave increases, the wave grows taller until eventually it topples over or *breaks*. It is these waves and *breakers* that are responsible for the many processes of erosion and deposition that are typical of every shoreline and beach.

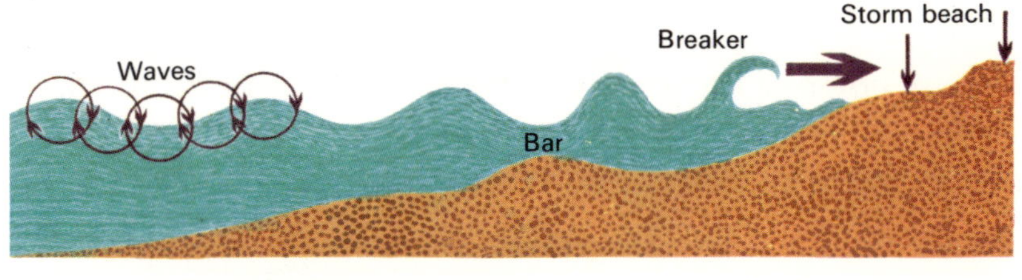

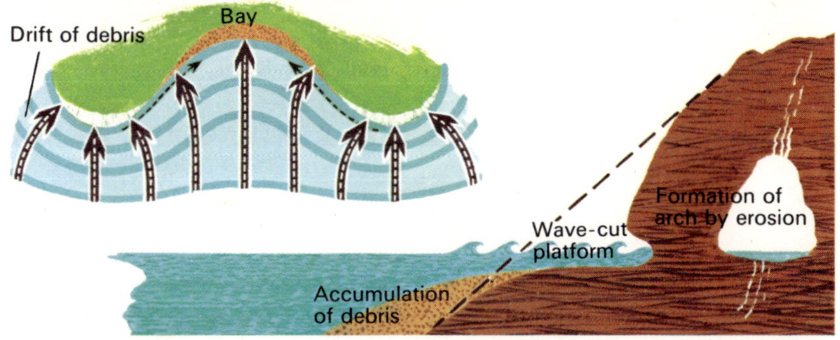

Left:
These diagrams show the movement of waves at sea and the way in which they break as they approach the shore. Waves may curve to erode a headland, and they can cause considerable erosion.

What are groynes?

When you have visited the seaside, you may have seen what look like garden fences running straight down the beach into the sea, and then coming to a halt. But they do not seem to be fencing anything off, do they? These barriers are called groynes. If you look more closely at them you will probably find that they are covered with various types of seaweeds and barnacles (small limpet-like shells but with a hole in the top). More important than this, however, you should be able to see that the sand or shingle has built up higher on one side of the groyne than the other. In fact, on the south coast of Britain, it is very likely that the sand will be higher on the western side, and on the east coast it will probably be higher on the north of the groyne. Why should this be happening? And why are the groynes there in the first place?

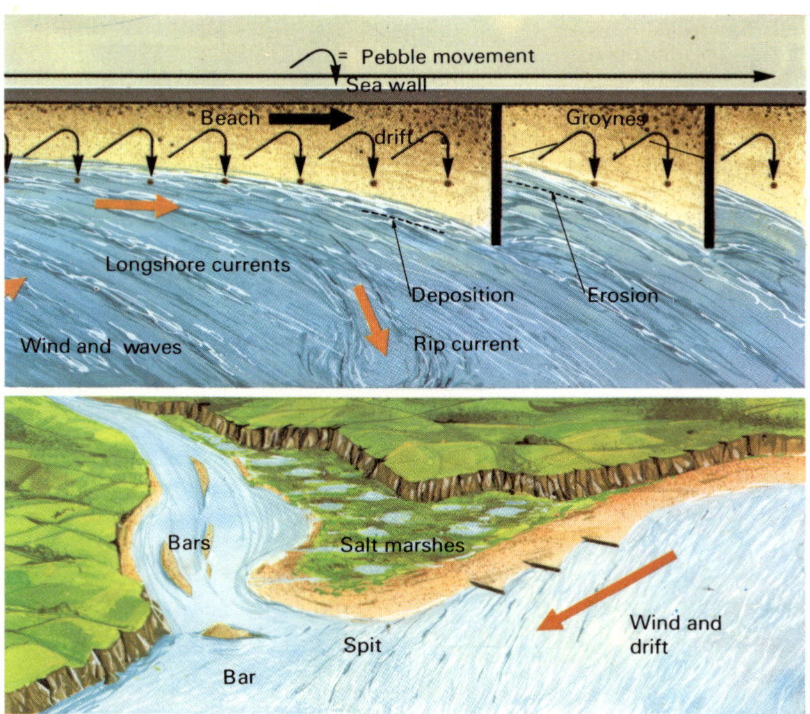

Right:
Groynes prevent sand and shingle from being transported very far by the action of longshore drift. Longshore drift may cause a spit to form at an estuary and salt marshes then develop behind it.

We have already explained how waves are formed by the wind and how these waves break when they reach the shore. Although it is only the wave shape that moves in the open sea, on the shore the breaking waves of water tend to have a motion shorewards. This motion is called the *swash*. But as you will have seen, after the swash up the beach, the water runs back again – this is called the *backwash*. If the force of the swash is greater than the force of the backwash, the

Right:
When the amount of sediment that is being carried by a river is greater than the amount that can be removed by longshore drift and other currents, a delta may form. The inset shows how the sediments in a delta build up.

waves are called *constructive*, and the sand, mud, and shingle that may be carried with the wave tends to be moved in the direction of the wave. When the reverse occurs the waves are called *destructive*. Generally, constructive waves are formed in quiet weather conditions and destructive waves during storms.

Generally, the waves do not come straight up the beach in Britain. This is because the wind most commonly blows from the west, that is, the *prevailing* winds are the westerlies. Therefore, on the south coast, for example, the waves travel up the beach and along from west to east. The sand and shingle is carried diagonally up the beach by the swash and directly down the beach by the backwash. This means that sand and shingle tend to move from west to east. What would happen if this were allowed to carry on unchecked? Of course, the beaches in the west would be washed away and those in the east built up. To prevent this the groynes are erected and the sand builds up on one side. Sometimes, the movement of shingle and sand may build up material into bars and spits. This is called *longshore drift*, and a good example of this is Chesil Beach near Weymouth in Dorset, England.

What are the main types of coasts?

In 1952, a scientist called H. Valentin grouped all the coasts of the world into two main divisions. He called these two types *advancing coasts* and *retreating coasts*. Advancing coasts may be the result of the uplift of the coastal land or they may be built outwards by deposition. Retreating coasts occur when the land is sinking or when it is being eroded by the force of the waves. Both these processes may be taking place at the same time on many beaches, and it is only by carefully observing an area over a period of many years that scientists are able to tell which process is the more important in that area.

Right:
A raised beach like this one is typical of the Isle of Arran.

Right:
A coast line showing the features of bays and headlands characteristic of submergence.

Far right:
An emerging coastline.

There are many features associated with these two main types of coastline which enable you to recognize them quite easily. For example, if the sea-level rises or the land sinks, it is clear that many of the features that were present on the land and formed by terrestrial processes will become flooded by the sea and drowned. If it was a hilly area before the sea advanced, you might expect to see bays, estuaries, rias (drowned river valleys), and gulfs separated by projecting areas of dry land called headlands and peninsulas. The *fjords* of Scandinavia are typical of submergence.

If you were to go to many parts of the north-west coast of Scotland, you might notice a different situation, however. On the Isle of Arran for example, behind the present beaches you might be able to recognize what appear to be other typical sea beaches. These beaches are higher than the level of the highest seas there today. They are called *raised beaches* and are old sea beaches which have been raised above the present sea level. Sometimes, as many as four levels of raised beaches can be seen.

Right:
The Temple of Serapis, near Naples, once stood on dry land. It was then submerged by the advancing sea but now the sea has fallen again and the borings made by marine worms can be seen in the columns.

What causes the changes of sea-level?

We have seen in previous questions how great sheets of ice have extended over many areas of the Earth's surface at certain times during geologic history called Ice Ages. The water which made up these icy masses came from the seas so that the sea-level fell, but at the same time the weight of the ice pressed down upon the Earth and when the ice retreated the land bounced back up again. The areas of the Baltic still seem to be rising from the melting of the ice from the last Ice Age. Local sea-level changes may occur for many other reasons such as earth movements.

What can you find on the beach?

Some of you reading this book will live by the seaside, and may have first-hand knowledge of the treasures that can be discovered by hunting along the beach, or by beachcombing, as it is commonly called. In Britain, people can visit the coast easily because Britain is quite a small island. In other parts of the world it is far more difficult.

We have already talked about the main types of coast line and classified them according to whether they are submergent or emergent. But for the beachcomber, a more useful division might be between those coasts with a shore and beach and those with the sea approaching the land at a steep cliff. Obviously, beachcombing along a cliff coast is not a possibility. Even among shores with a more gentle approach, there are differences. For example, there are beaches blessed with stretches of golden or silvery sands, such as the beach at Perranporth in Cornwall, England; then there are those with pebbly beaches such as the famous spit of Chesil Beach at Weymouth in Dorset; there are others that have more rocky beaches such as those found along the coast of Pembrokeshire in Wales. This is a very rough and ready classification but is useful because all of these types can offer something different to the keen observer.

Let us look first at sandy beaches. Obviously, there is the sand, and although you may not notice it at first, with practice you will come to realize that there are various colours of sand depending upon the mineral grains which make it up. Secondly, there are things that live and grow on the sand (or in it) and thirdly there is the flotsam and jetsam washed on to the shore, ranging from bits of wrecked ships, to plastic bottles, dead fish and seabirds.

Right:
Many interesting items can be cast up on a beach.

Left:
Cockling is often a worth-while occupation but should never be done near a sewage outlet.

Plants that might be expected to live on sandy beaches are marram grass among the dunes, yellow horned poppy if it is a little more pebbly, or sea aster slightly further from the shore. You might find animals like the lugworm or double shelled molluscs, like the cockle living under the sand. You could also find the egg cases of whelks and many other creatures.

On pebbly beaches like Chesil Beach, you might notice that the pebbles change in size as you walk from one end of the beach to the other. This is a result of longshore drift that has already been mentioned.

Rocky beaches offer a great deal of interest to the keen observer. There are the rock pools to be explored which are left behind when the tide goes out. You might see the tentacles of the sea anemone waving in the currents, waiting to sting an unwary shrimp, fish or crab. Look out for limpets and barnacles attached to the rocks. Limpets are conically shaped; barnacles are smaller and have a hole in the top. We should not forget the many varieties of seaweed on all these beaches. The important thing is to look, ask and learn!

What can you do with pebbles from the beach?

At first, you might think a visit to a pebbly beach is less exciting or interesting than going to a beach with golden sands. After all, sand is much more comfortable to lie on and it does not hurt your feet when you go for a paddle or a swim. On the other hand pebbles can be such fun once you know what to do with them.

Even if you do prefer sandy shores, you have probably visited a shingle beach and noticed the glorious array of colours and strange and wonderful shapes that pebbles sometimes present. Have you taken home a flint pebble with a hole through the middle or been lucky enough to find a beautifully banded agate? The shapes and colours of the stones you find depend upon what the pebble is made of and how long it has withstood the battering of the sea. You may have noticed that wet pebbles near the sea look much more attractive than those found further up the beach that have had time to dry in the sun. This is not because there is any difference in the stones but simply that pebbles appear more shiny and brightly coloured when they are wet. If you like the look of wet stones you can preserve their appearance by coating them in varnish. Lapidary shops can sell you the best types of varnish for this purpose.

There is another way of improving the look of pebbles from the beach which more and more people are finding to be a fascinating hobby. Unfortunately, it does mean that you will probably have to buy some tools for the job. In particular, you will need a tumble polisher which consists of a hollow plastic drum. With the aid of water at first, and then

Right:
We have already mentioned longshore drift (*see* page 67). It can lead to a sorting of pebbles as you can see in this illustration.

Left and below:
You can see how pebbles from the beach can be polished and made into fashionable jewellery.

various abrasive powders, your stones will be smoothed and polished as they are rotated by an electric motor. We have not tried it, but we expect it would be possible to make a polisher with the help of, say a Meccano outfit and a little imagination.

When you are satisfied that your stones are looking really pleasing, there are various ways of showing them off to their best advantage. You could arrange them skilfully in a glass cabinet perhaps indicating on a label what the stone is and when and where it was found. Or even more fun, you could make them into various types of jewellery by glueing them to metal mounts that you can buy from a lapidary stockist. Once again, the motto is to be observant, learn, and use your imagination, and the beach can provide hours of interest.

Lakes – what are they? How do they form?

A lake is usually the name given to any large body of water which is trapped within an area of land. In fact, some ponds and pools could be called lakes, and even inland seas, such as the Caspian sea, are strictly lakes. Nor should we forget the cut off meander loops called oxbow lakes.

It is interesting to note that lakes are characteristic of areas that have been subjected to glaciation. The English Lake District provides us with a classic example. It may surprise you to learn that the majority of the world's lakes today owe their existence to glacial action. In areas such as the Lake District, less resistant rocks are eroded by ice action more easily than harder rocks, so that large numbers of hollows are formed. Further deepening by the glacier also provides natural lake basins such as the well-known corrie lakes.

A lake can only continue to exist if it is being refilled with water from rivers or some other source at a greater rate than water is being lost from the lake. Huge quantities of water are lost by evaporation, particularly from the surface of larger lakes. In parts of North America where great dams

Right:
Faulting can give rise to lakes. This is found in the East African Rift System as shown on the map.

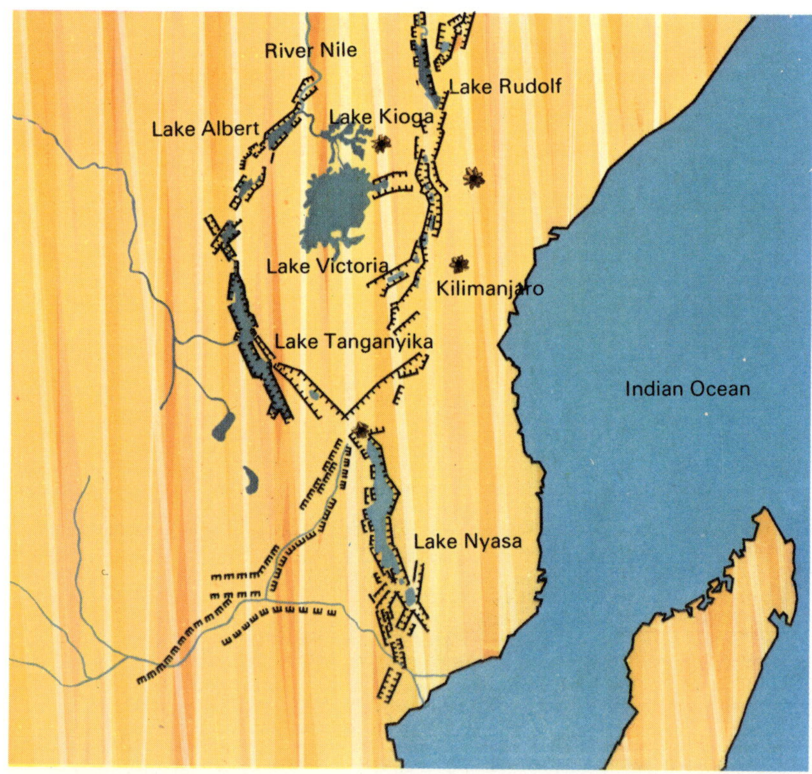

Right:
A familiar lake scene.

Far right:
A desert lake.

have created vast artificial lakes for irrigation and hydro-electric purposes, the increased water losses are considerably increasing the salts in the water. Other losses of water may occur through leaking away at the sides and floor of the basin.

An important factor in the life of a lake is the rate at which sediment accumulates. It is not difficult to understand that lakes fed by silt-laden rivers quickly become filled with sediment. Taking all these factors into consideration, you can see that in the context of the history of the Earth, lakes cannot be thought of as very lasting features of the scenery.

In addition to glacial lakes, there are a number of other types. Some have resulted from various movements of the Earth's crust. For example, the world's deepest lake, Lake Baikal in the USSR, is situated in the Baikal Rift System. Other lakes are associated with volcanic activity. The Sea of Galilee was formed by a lava flow blocking a valley that was already present. River valleys may also be blocked by land-sliding or glacial deposits so that a lake builds up behind the barrier. A rather more unusual origin for some of the world's lakes is the impact of large meteorites. When these meteorites strike the Earth's surface, the impact creates a crater which in most cases remain dry, but occasionally they become filled to form such lakes as Lake Bosumtwi in Africa.

Right:
The lake at Cader Idris, Wales.

Can rocks bend?

When you pick up a pebble on the beach, it is hard to imagine that this solid material rock, can be bent just as though it were toothpaste. Throughout the world, however, you can find large areas of exposed rock that clearly show the evidence of bending. It is also important to realize that it is the bending and upheaval of the surface rocks which give rise to the basic structure from which the landscape is moulded. Some of the *folds* which have been formed in beds of rock are so large that you cannot see their complete shape. Their existence can be proved by other methods, as we shall see later. You will also see later how rocks called sedimentary rocks are deposited in beds that are almost horizontal, but they can be twisted into remarkable shapes and even turned upside down by later Earth movements.

Geologists try to simplify the business of examining folded rocks by giving characteristic forms names. Take a piece of paper to help you learn some of the folds that can be seen in nature. First of all, bend the sheet into a gentle curve and hold it so that the bend is at the top as you look at it. When rocks are bent into this shape it is called an *antiform*.

Now, instead of using just one sheet, do the same thing with a number of sheets held together. Try to imagine that

Right:
This illustration shows a part of a fold. The *dip* of the rocks is important to the geologist who is trying to work out the structure in an area. If you were to walk along the *strike* of a bed of rocks you would be walking along the same bed continuously.

Centre right:
A clinometer is an instrument for measuring the dip of a bed of rocks.

Far right:
This illustrates how a fold may be fractured.

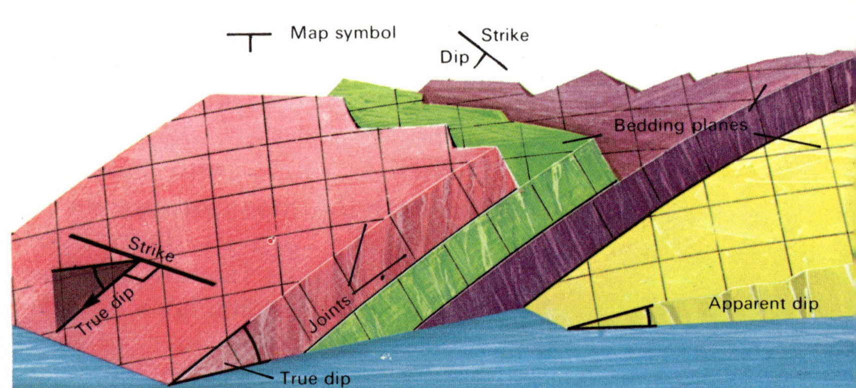

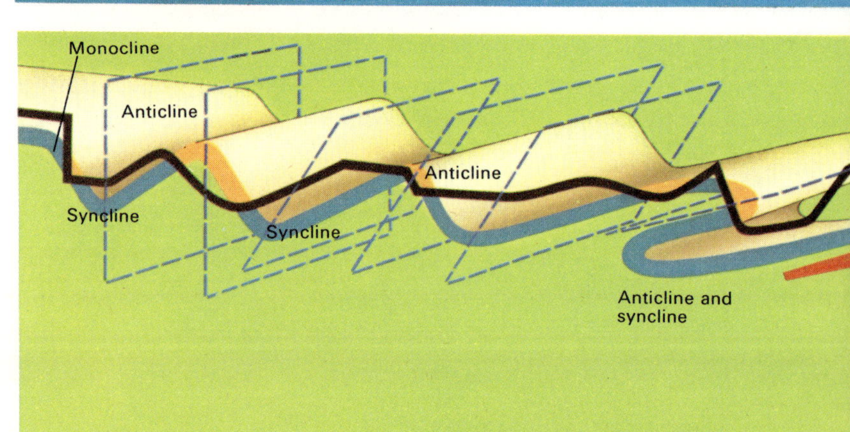

each sheet of paper is a bed of rock and that the bottom sheet was the first bed to be laid down, in other words the oldest. Now hold the fold in one hand and cut off the top of the curve with scissors. You will see that the sheet that was at the bottom is now in two halves at the centre of the sheets of paper. When this happens to folded rocks, and the oldest beds occur in the centre of an antiform, it is called an *anticline*. Sometimes if there is more than one period of movement, what was at first an anticline can be turned upside down, and the situation becomes quite confusing.

Of course, beds of rock in the Earth's crust are rather longer than your piece of paper and you would not expect to find just one fold. Take the sheet of paper and put two gentle folds in it. You will see that one has its bend at the top, the antiform, but the other has its bend at the bottom; this called a *synform*. When there are younger rocks in the centre of the synform it is called a *syncline*.

With many periods of upheaval, then, you can see that there could be synformal anticlines and antiformal synclines. Sometimes, rocks are bent into *dome* and *basin* shapes.

Above:
Scientists have tried to reproduce the folds that can be seen in rocks by compressing layers of clay.

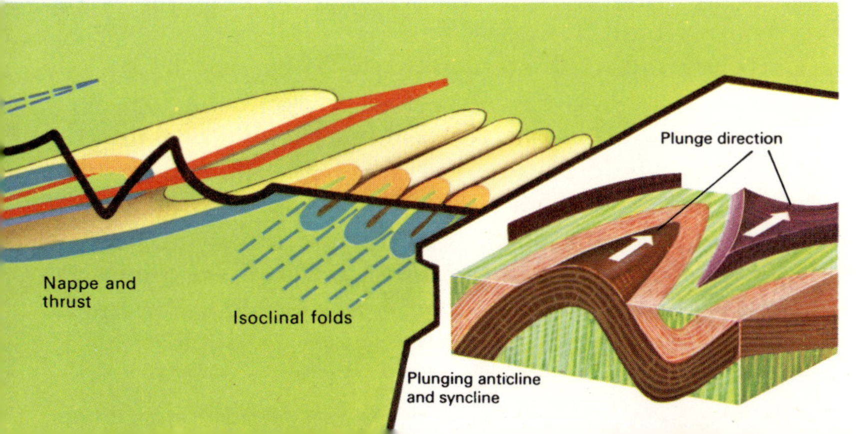

Left:
This diagram shows some of the different types of folds.

Can rocks break?

You have seen that the enormous forces locked within the Earth are enough to bend huge masses of rock. Sometimes the rocks may break and great blocks may move quite large distances. This is known as *faulting*. A famous fault in Britain is the Great Glen Fault running across Scotland from the Island of Mull in the south west to the Moray Firth in the north east. This is almost as though Scotland had broken

Right:
Some of the major faults and folds of Britain.

into two pieces. In fact, along this line the northern part has moved south westwards relative to the southern half by a distance of as much as 105 kilometres. This type of fault is called a *tear fault*, and they are very important on a global scale as we shall see later. A well-known example is the San Andreas fault in California.

There are many different types of faults, and faults, like folds have their own special names. If you could stretch a bed of rock until it broke, you would probably find that the way it broke would be usually called a *normal fault*. In other words, the surface where the break took place would be quite steep, and one block would move downwards along this plane with respect to the other so that if the plane was inclined to the left the right-hand block would be the lower.

Normal faults, then are usually associated with crustal stretching.

On the other hand *reverse faults* are associated with crustal shortening. In the above situation, a reverse fault would have the lower block on the left. In some cases, reverse faults occur where the fault plane is almost horizontal. This type of fault is called a *thrust*.

As you can see, on any given fault plane there can be vertical movement, that is, up and down, and this is called the *throw* of the fault. There can also be horizontal movement, that is, from side to side, and this is called the *heave* of the fault. Many of these strange terms that have become associated with faulting had their origins in the coal mines of the north of England where faults were particularly important. One way of discovering the presence of a fault is by noticing that there is a sudden change in the beds of rock from one side of the fault to the other. For example, on one side of a fault there might be Carboniferous limestones and on the other older Devonian sandstones. This would mean that the younger limestones had dropped down in relation to the sandstones. Other miners' terms associated with faulting include the word *hade*. The fault plane of a normal fault, for example, slopes towards the fallen block of rock, that is, it 'hades to the downthrow'.

Right:
Some typical types of fault movements.

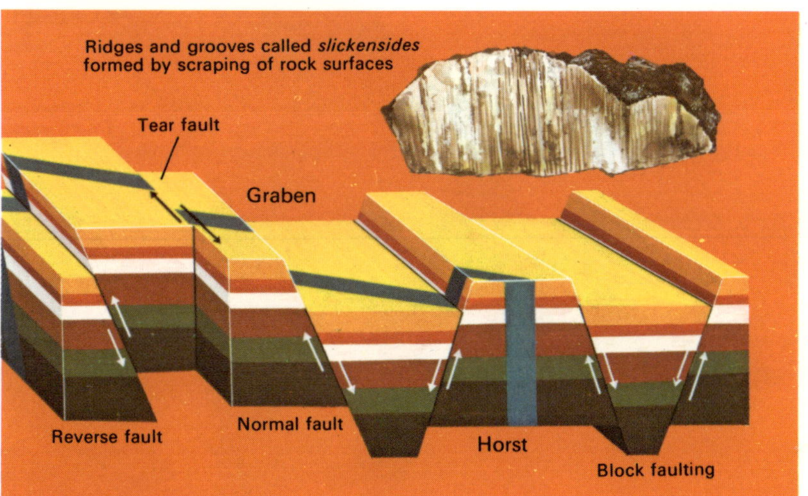

Where there are combinations of reverse and normal faults, whole blocks of rocks may be moved up or down. A block which has moved up is called a *horst*, and a block that has moved down is called a *graben*.

What do you know about

The Meaning of Fossils

What is evolution?

Above:
Charles Darwin.

To answer the question very simply, evolution is the process by which living things on our planet Earth have changed and developed from the algae-like forms of which there is evidence in ancient rocks, into the many varied plants and animals which we can see around us today. But if we were to leave it at this simple explanation, there would be much that would remain unsaid.

As we have already explained, life first seems to have shown itself on Earth some 3500 million years ago, and if we are to believe what the modern theories of evolution tell us, it has taken all that time for the modern creatures to evolve. It is written in the Christian Bible that God created the Earth and all its inhabitants in six days and nights and on the seventh day He rested. Many books have been written, however, by men of religion, by men of science, and by men of both religion and science discussing the merits of the ideas of evolution and of religion. In this question, we shall be concerned only with the evidence that can be drawn from such sources as the fossil record, and Charles Darwin's historic voyage in HMS *Beagle*.

In the 1600s Archbishop Ussher calculated that the Earth was formed in the year 4004 BC and this view held fast until the late 1700s when the science of geology was taken more and more seriously. At this time a scientist called Hutton stated his *Principle of Uniformitarianism*. He said that all the processes of the Earth took place very slowly; this developed into the idea that the present is the key to the past. In other words, if you look at a process's results on Earth today, such as the kind of muds and silts that are deposited in a modern estuary, and you then find similar sediments in an ancient rock, these sediments must have been formed by an estuary. From these ideas it was realized that the Earth must be very old indeed. Then William Smith at the turn of the 1700s to

Right and far right:
Pigeon fanciers have been able to exploit the processes of evolution by selecting individuals with particular characteristics and breeding from them so that eventually different types emerge like the ones seen here.

Right:
The very early stages in the lives of very different animals are quite similar. It is thought that this is because the animals had a common ancestor.

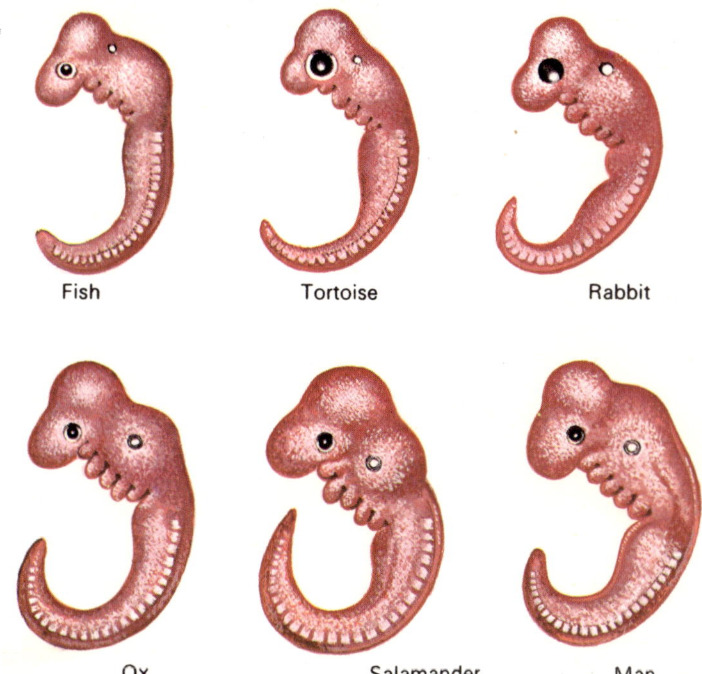

the 1800s showed how it could be proved that layers of rock were laid down one on top of the other over long periods. From this point the value of fossils was recognized because it could be seen how the older fossils were very simple, becoming more and more complex in the younger rocks.

Charles Darwin and evolution go hand in hand. Darwin travelled to the Galapagos Islands and noticed that birds which were all clearly finches had adapted by evolution to live in a variety of ways. Elsewhere, other species of birds had filled all the available ecological niches. He said that this had occurred because the islands were isolated and the finch had evolved to fill these gaps.

chalk

sandstone

limestone

shale

Above:
Some typical sedimentary rocks in which fossils may be found.

Right:
An animal may be fossilized like this if the conditions are suitable.

What is a fossil?

Fossils are any remains of a once living plant or animal that have been preserved by a variety of ways. The trace may be in the form of a complete animal such as the mammoths which were found by Russian scientists frozen into the ice of Siberia. It was said that the meat on the animals was still fresh enough to eat, and indeed Russian geologists were supposed to have dined on mammoth steaks. This form of preservation is very unusual. A more likely way is for the hard parts of the animal to be preserved as a fossil. The fossil is formed when the animal dies because the hard parts are buried by sediments which in turn become compacted into rock, leaving the fossil contained within. Insects have commonly been found preserved in amber which is fossil tree sap. The sap probably ran down the trunk of the tree trapping the insect so that when the sap hardened into the beautiful yellowish brown amber it was preserved complete.

Plants may be found preserved as black traces, particularly in rocks called shales. You may find traces of leaves and stems in this way. More recently it has been discovered that the pollen which all flowering plants release for fertilizing their seeds has been preserved in quite large amounts. In fact, the science of palynology (the study of fossil pollen) as it is called, has become very important in the search for oil.

Animal trails may also become fossils. Trails of worms moving along the sea bed have been found which have become 'frozen' into what has become a rock. Footprints of the long dead dinosaurs are not unusual. In the same way that modern chickens eat grit and retain it in their crops to

living animal

hard parts of dead animal

rapid burial

impregnation by minerals from sediments

aid digestion, dinosaurs swallowed quite large stones and they can be recognized by scientists called *palaeontologists*, who study all forms of life in former geological periods through fossils.

Which fossils are you most likely to find?

The animal kingdom is divided into those animals which have backbones, *vertebrates*, and those animals which do not have backbones, *invertebrates*. In terms of the numbers of animals which have ever lived throughout the history of the Earth, the invertebrates are much more common. Many invertebrates had external shells made of chalk-like materials to protect them. Others had internal skeletons made of similar substances.

We mentioned in the previous question that it was usually the hard parts of animals that were preserved. Therefore, you are most likely to find fossils of animals which were numerically common, widely distributed, possessed hard parts, and which lived in conditions suitable for their preservation. In other words you are most likely to find fossil invertebrates.

Right:
A mammoth trapped in soft mud may be fossilized and its bones rediscovered millions of years later.

What are the geologic column and the fossil record?

As you know the Earth is about 4500 million years old. Clearly, this is an immensely long period of time, difficult for us with our life span of about 70 years to imagine. To make consideration of the history of the Earth a little easier, this huge time scale has been divided up into smaller units, and then further divided into still smaller units. The largest unit of time that we use is called an *era*. There are six of these eras, but they do not by any means represent equal lengths of time. They are the Eozoic, Archaeozoic, Proterozoic, Palaeozoic, Mesozoic, and Caenozoic. Palaeozoic, for example, means ancient life, and the rest have equivalent meanings. During any given period of time, however, beds of rocks will have been laid down somewhere around the globe and these must be given names too. The strata or beds of rock deposited during an era are referred to as a Group.

Eras are divided into Periods and the Groups into Systems and it is the Periods and Systems which make up the geologic column shown in the picture. Sometimes confusion arises between time and rock units. If you wish to divide say, the Devonian period you might refer to the divisions as Early, Middle, or Late Devonian time, whereas the equivalent rock units would be the Lower, Middle and Upper Devonian. It is important that these terms should not be mixed up, although they often are, even by experienced geologists.

As you can see from the picture, the last 600 million years only comprise three Eras. This is only a tiny period compared with the 4000 million years that went before, which are usually referred to loosely as the Precambrian, that is, the period of time before the Cambrian, the lowest Period in the Phanerozoic time scale.

At first, it was thought that the first life appeared on Earth at the beginning of Cambrian time because it was in rocks

Right:
Fossils may be used to date rocks of different types in areas some distance apart.

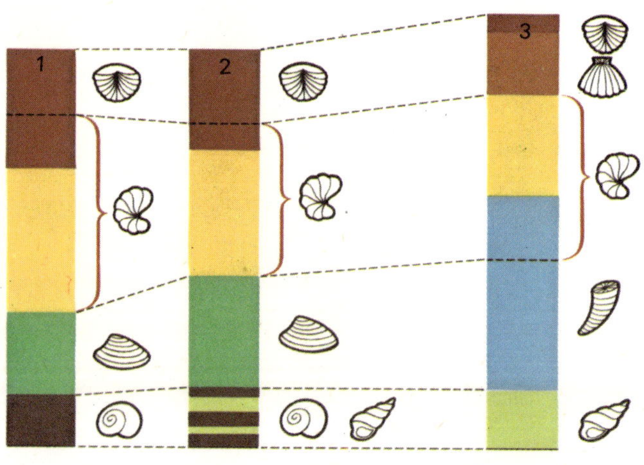

Right:
The geologic column and fossil record.

described as Cambrian in Wales that fossils were first found. Later, however, it was discovered that the first life forms may have occurred at least 3000 million years ago, but still the most detailed fossil record exists for the last 600 million years.

The boundaries between the time Periods or rock systems are often established on the appearance of some new form of life. It may be a completely new type of life form or just a distinctive species of an animal or plant form that had existed earlier. In fact, the availability of the fossil plays an important part too. For example, although graptolites had existed in Cambrian times, the Ordovician period is usually subdivided on the basis of the species of these animals.

What was the first life on Earth?

Above:
The Precambrian fossil called *Charnia* discovered in Charnwood Forest, Leicestershire, England. It may have been an early seaweed.

Until quite recently it was thought that there were no abundant life forms on this planet until about 600 million years ago. When we remember that the Earth itself is at least 4500 million years old, it does seem to have taken a very long time before any animals or plants appeared. Then, geologically speaking, they must have developed very quickly indeed. Probably the reason why 600 million years was thought to be the date of the earliest common life was because a great deal of the pioneer work in geology in the 1800s was done in Britain. Apart from some recently discovered traces in older rocks of Scotland and Charnwood Forest, England, the oldest fossils occur in rocks dating from a period called the Cambrian (*Cambria* Roman name for Wales) found first in Wales and then in Scotland. In these rocks fossils of shellfish called brachiopods, insect-like, sea-dwelling trilobites, cockle-like shellfish, and one or two other groups of animals are all to be found in abundance.

It was obvious, however, from the work of many geologists that these Cambrian rocks were not the oldest in Britain. Unfortunately almost all the earlier beds had been severely altered by heat and pressure as a result of earth movements so that even if there were any traces of life they would have been lucky to survive for the workers of the last century to find. All these older rocks were lumped together, and the period in which they were deposited was called the Precambrian (before the Cambrian).

Right:
These Precambrian fossils can be found in the ancient rocks of South Africa and Australia.

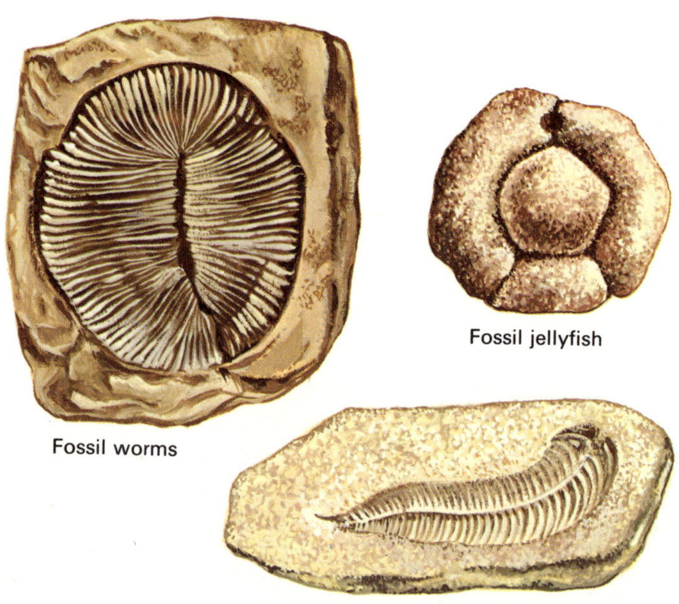

Of course, if many different types of animals had developed by the Cambrian period, it seems reasonable to suppose that they could not all have suddenly arrived on the

Spriggina

Above:
This fossil, called *Spriggina*, may have been an early worm.

Right:
This picture indicates some of the animals that may have existed rather later in the Earth's history, in the Middle Cambrian—a mere 550 million years ago. By this time several animal species had evolved.

scene. This suggested that although few traces had been definitely identified as belonging to more primitive life, there must have been some early life forms developing slowly during the millions of years of the Precambrian.

Later more and more people in countries outside Britain became interested in the geology of their own countries and it was not long before traces of very primitive worms, sponges and other very early invertebrates were found in Precambrian rocks of North America, Australia, and southern Africa. In 1964, a scientist called J. G. Ramsay claimed to have found structures which resemble the algae you might find in a pond, in a series of rocks which have been dated at 3500 million years old. In fact, associations of algae and other life forms, called stromatolites, are the most widespread Precambrian fossils. You can see then that the fossils which you may be able to find quite easily on beaches or in chalk quarries are very old by our standards but newcomers by the standards of Earth's time.

What are the Brachiopoda?

Strophomena Ordovician

Schizophoria Devonian – Permian

Chonetes Devonian

Conchidium Silurian – Devonian

Above:
Some brachiopods ranging from 500 to 200 million years old.

Above right:
The parts of a brachiopod shell.

The Brachiopoda comprise a group of animals which were found in abundance in rocks of Cambrian age some 600 million years old. At least one member of this group is still to be found today. They are simple animals which range in size from a few millimetres to a maximum of 30 centimetres across. They are protected by a hard shell which is made up of two parts, known as valves, rather like that of a modern cockle. Almost all these animals lived in the sea, although the modern member of the group, *Lingula*, tends to live in a burrow in more brackish water. Lingula existed in the Cambrian and it is interesting to note that in 600 million years this animal has hardly changed at all. Even though it is very simple, it has really been quite successful in the battle for survival.

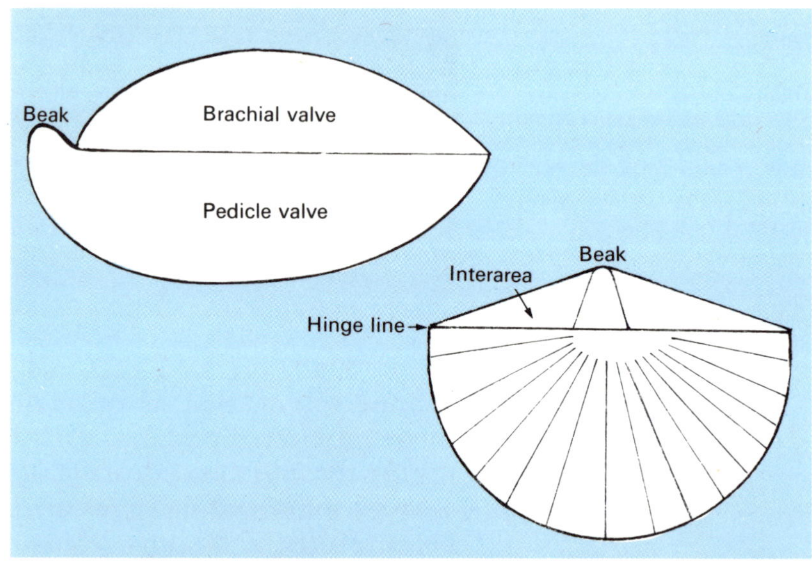

From the type of rocks in which brachiopod fossils have been found, it can be seen that they lived in quite shallow water, certainly no more than 15 to 30 metres below low water mark. The shells of some of the primitive members of the group were made of chitin, which you will see also composed the external skeleton of the trilobites that we have already discussed. The later brachiopods' shells were composed of calcite which has the same chemical formula as the chalk you find in many cliffs.

All these animals fed by extracting tiny food particles from the water. Animals which feed in this way are known as microphyllous feeders. The feeding method was really very simple. A hard loop of skeletal material inside the shell was

Right:
These brachiopods are examples from more recent periods.

covered in tiny hair-like structures called cilia. The cilia were covered in sticky mucus. The animals waved these hairs in a rhythmic fashion, creating a current of water through their shells. The water carried the oxygen needed for the animal to breathe and tiny pieces of food which would be trapped on the sticky mucus of the cilia. The moving stream of mucus then carried the food into the brachiopod's mouth.

Brachiopods were attached to the sea floor by a strong stalk called a pedicle. Although they could not move from place to place, they were able to rotate their shells on this stalk to make the most of the food and oxygen in the water.

When the animals are born, they do not have hard shells, nor are they attached to the sea floor. This means that the tiny young are carried great distances by the tides and currents of the sea. As a result they are able to spread over a wide area quite quickly. It is at this stage that these tiny creatures are most likely to be eaten by other animals in the sea. After some time as these free-floating forms, the first shell is produced and the brachiopod settles to the sea floor. They then have only twenty-four hours to find a suitable place to live or they will die. It is interesting to note that there are other hard-shelled animals which have this free-floating stage in their life cycle.

Almost every group of animals, including brachiopods, have developed throughout geologic time with the intention of increasing the efficiency of their breathing and feeding.

Brachiopods are widespread in rocks of a suitable age. As they changed quite rapidly, geologists find them useful to date the ages of the rocks where they are found.

Which animals were the dustbins of the ancient seas?

From the beginning of the Cambrian period in Earth's history until the end of the period called the Permian, some 375 million years later, a group of animals called Trilobites carried out the vital role of scavenging. This means that trilobites disposed of all the waste products which may otherwise have built up in much the same way as vultures do today.

Trilobites are included in a much larger group of animals called Arthropoda (meaning jointed legs). Today, they are represented by such creatures as the insects and the spiders. It is worth noting that spiders are not insects; remember that spiders have eight legs and insects have six.

Arthropods have always played a very important part in the complicated web of life on Earth. Even today, there are more kinds of insects in far greater numbers than any other animal. You have probably been told about the vast swarms of locusts which sweep across Africa eating all of the crops in their path. These swarms may be made up of many millions of individuals.

The word 'trilobite' means 'having three lobes or parts'. If you are lucky enough to find a complete fossil of one of these most interesting of life forms, you might be tempted to think that these lobes were the head, the body, and the tail,

Right:
These fossils are examples of early forms of trilobites.

Olenellus Cambrian

Paedeumias Cambrian

Paradoxides Cambrian

Bumastus Ordovician – Silurian

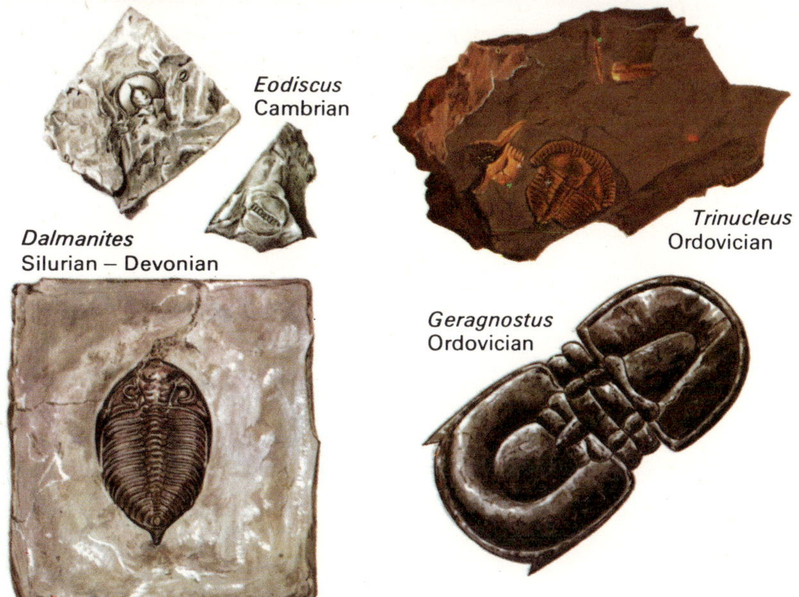

Right:
Trilobites became extinct about 225 million years ago.

or the cephalon, the thorax, and the pygidium as they are more scientifically called. In fact the parts referred to are the lobes which run the length of the animal. They are called the axis and the two pleural lobes.

As we have already mentioned, trilobites were already quite advanced by the beginning of the Cambrian and from there they developed to live in a variety of ways, but they all lived in the sea. It seems that most of them scavenged on the sea bed, but some kinds were free swimming and others lived by burrowing in the soft silt and mud. If you find any of these fossils, try to work out for yourself the conditions in which they lived. A clue is that not all of them had eyes.

Trilobites had a hard outer covering of chitin and in order to grow, they were obliged to shed it from time to time. They probably laid eggs but no fossil eggs have ever been found.

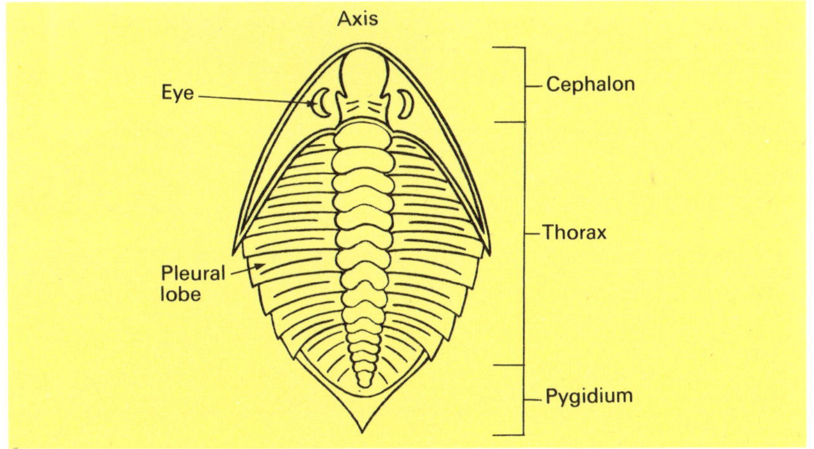

Right:
The structure of a trilobite's body.

Which extinct group of animals' name means 'stone writing'?

The word 'extinct' describes a group of animals which is no longer living. The name of this particular group is *Graptolites* from the Greek *graptos* meaning writing and *lithos* meaning stone. The first traces of this group are to be found in rocks deposited during the last part of the Cambrian period in Earth's history. The reason that they were given this rather curious name is because of the form in which they are usually found. If you were to split open rocks of the correct age you would see what look rather like tiny, black, shiny saw blades on the surface of the rock. These traces usually look like some form of ancient writing.

In fact, these dark traces are the remains of graptolite skeletons (which were originally made of a material like chitin) that have been flattened and altered to carbon. As you know, carbon is the main component of coal and is also to be found in the so-called 'lead' in your pencils.

Each branch that we see is actually a colony of these tiny animals, the individuals living in cups called *thecae* which have been flattened into the serrations along the branches. A typical example of one of these colonies is called *Didymo-graptus*. This consists of two branches emerging from a growth point usually called a *sicula*. Each branch is called a *stipe*. The whole colony would only be about five centimetres long, and is often in the shape of a 'V'.

Right:
Graptolites are usually found as carbonized forms on the bedding planes of rocks.

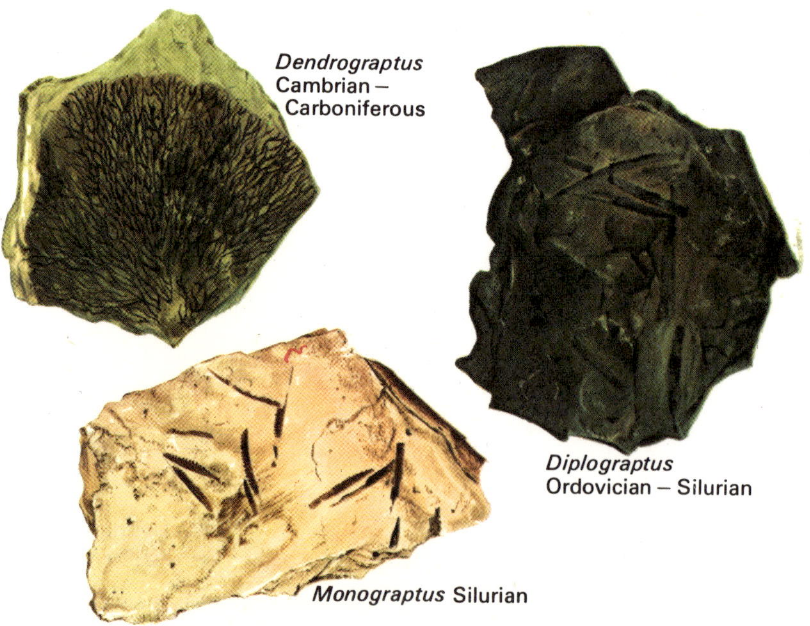

Tetragraptus Ordovician

Didymograptus Ordovician

Right:
Some more graptolites.

When they were first discovered, graptolites were thought to be plants, and it is easy to understand why, with the thecae looking like leaves on a stem. Then, some specimens were found that had escaped the flattening which is usual. It was then realized that the 'leaves' were cups and must have contained animals. It was suggested that they belonged to a group of animals called the *Coelenterata* which includes the beautiful corals found in warm, tropical seas today. This view was held until 1937 when a Polish palaeontologist called Kozlowski began to investigate these strange colonial animals. He found that far from being related to corals, they were more akin to animals with backbones, including man himself. As you might expect, it took a long time for such an extraordinary suggestion to be accepted. Today, still more evidence has been found to support Kozlowski.

This group of animals developed very rapidly during the 150 to 200 million years in which they existed on this planet, and many different forms have been discovered. They were all colonial, however, and they all lived in the sea, perhaps attached to floating seaweed.

Palaeontologists often divide the graptolites into two main groups depending upon the form in which the colony occurs. The most primitive group is referred to as the dendroid graptolites. The most primitive of these have sixty-four branches or stipes gradually reducing to sixteen as they evolve. Those graptolites with eight to one stipes are referred to as graptoloids and such species as *Tetragraptus* illustrated.

Which are the most primitive molluscs?

The *Mollusca* or molluscs are second in the number of species to that other vast group of animals without backbones, the arthropods. Molluscs are, however, of more interest to geologists. Molluscs are adapted to live under most conditions that the Earth can offer and some members of the group can survive in very dry environments for quite long periods.

The most primitive molluscs are probably the group known as the *Gasteropoda* or gasteropods, meaning 'stomach footed'. Primitive in the sense used here, means that the gasteropods most nearly resemble the simplest mollusc that could possibly exist.

Right:
The structure of a gasteropod shell.

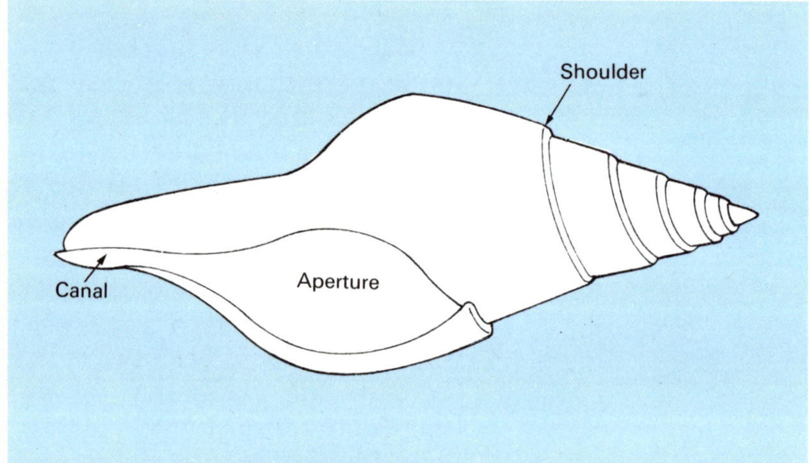

The animal itself consists of a small head containing the mouth which passes straight into its gut. The rest of the animal's bulk is mainly foot. It has no brain or central nervous system, but it does have a simple nerve ring. You have almost certainly seen slugs and snails in the garden or even in the house in certain weather conditions. Slugs and snails are gasteropods and members of this group are alive today in very large numbers; indeed they may be at their period of greatest development. The first gasteropods, however, appeared in the Cambrian.

Gasteropods may live in a very wide range of environments from dry land, like the slugs and snails already mentioned, to fresh water, the sea shore and even to the deep oceans. Some have protective shells and others do not. The ones that have shells are the most valuable to geologists because these hard parts can be fossilized.

Most gasteropods that have shells, have ones which are

Right:
Some primitive gasteropods.

twisted into some kind of spiral or flattened coil. This twisting of the shell mirrors (but not exactly) the twisting of the animal living inside it. The limpets that you can find fastened hard on to rocks on the sea shore are a modern exception.

Although gasteropods are fairly common in sedimentary rocks, they are not often used by geologists to date beds of rock. This is because many of the species of gasteropods have existed unchanged for quite long periods. Often, there are better fossils in the rocks in which they occur.

If you want to collect gasteropods for yourself they are quite common and may be readily identified with the aid of books.

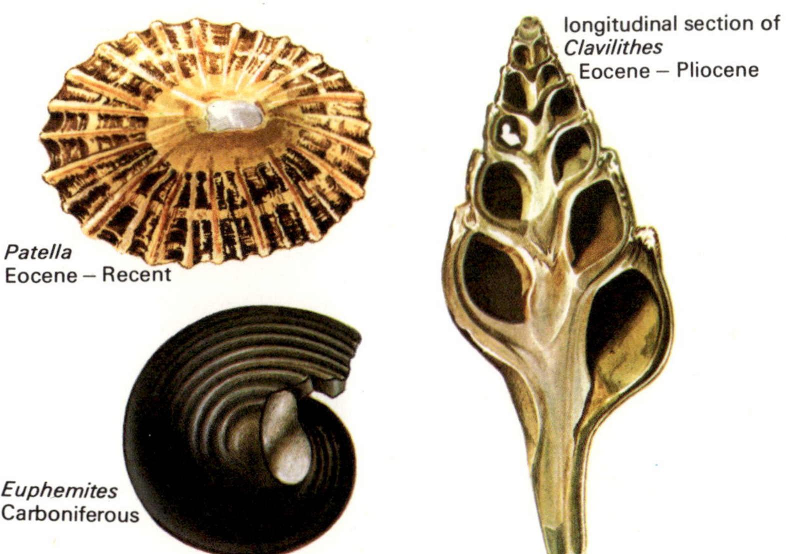

Right:
These are more advanced forms of gasteropods.

Which molluscs have a shell in two parts?

This group of invertebrate animals has been given a variety of names. In America they are usually called pelecypods, but in Britain and Europe they are called lamellibranchs or, more recently, bivalves. This last name is probably the most useful because it accurately describes the form of shell which protects these animals; it means two

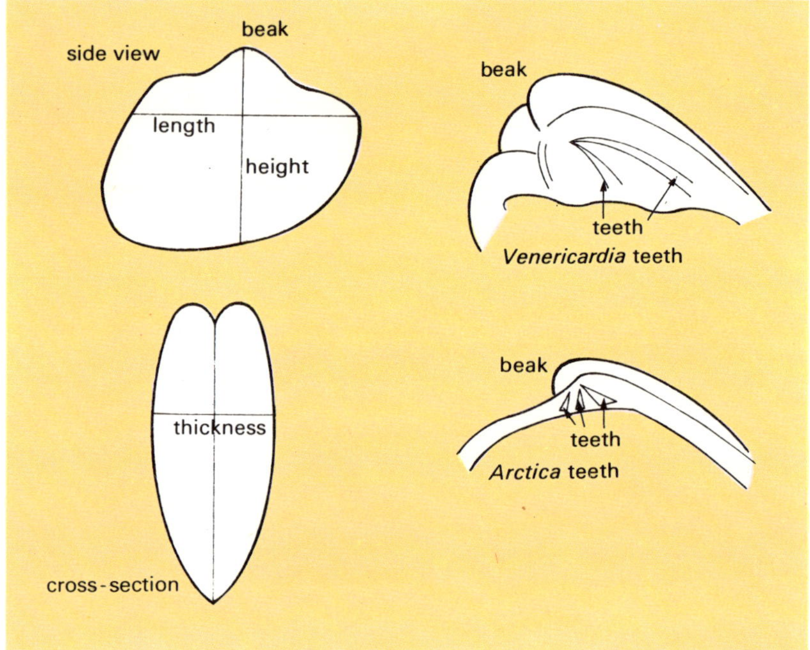

Right:
The structure of a bivalve shell.

valved or two shelled. Cockles are modern, familiar examples of this group. You may have found them buried in the sand or mud on the foreshore at low tide with only their siphons sometimes visible. If you have not seen the animals alive, you may have seen them in jars – without shells of course!

Bivalves never live on dry land; in fact, most members of the group live in a shallow water marine environment, but there are some that can survive in fresh water providing it is flowing. Unlike the gasteropods, they cannot survive in lakes and ponds.

Brachiopods, which we have already mentioned, are the other group of animals with a hard shell in two parts. Like the brachiopods, the bivalves are also microphyllous feeders – do you remember what this means? They live in the sediment in the sea floor, or attached to rocks, or on the sea bed. They then create a current of water through the shell from which they extract oxygen to breathe and food to eat.

When necessary the two halves of the shell can be pulled hard together by muscles called adductors. When these muscles are relaxed, the valves tend to fall open because of the pull exerted by the elastic band-like ligament at the point where the two halves of the shell are joined together. This is why it is more usual to find fossil bivalves in separate halves. When the animal dies, the shells naturally open and the force of the sea can easily break the remaining ligament before the shell is buried. If you find a shell that is open, but not separated, look at the point where they join. You will see tooth-like structures which help to keep the valves in the correct position. These teeth can help the geologist to identify fossil bivalves.

Once again, we know that bivalves are alive today, but they first appeared after the Cambrian at the beginning of the Ordovician period, some 500 million years ago.

Right and below: Many bivalves that can be found fossilized still live today.

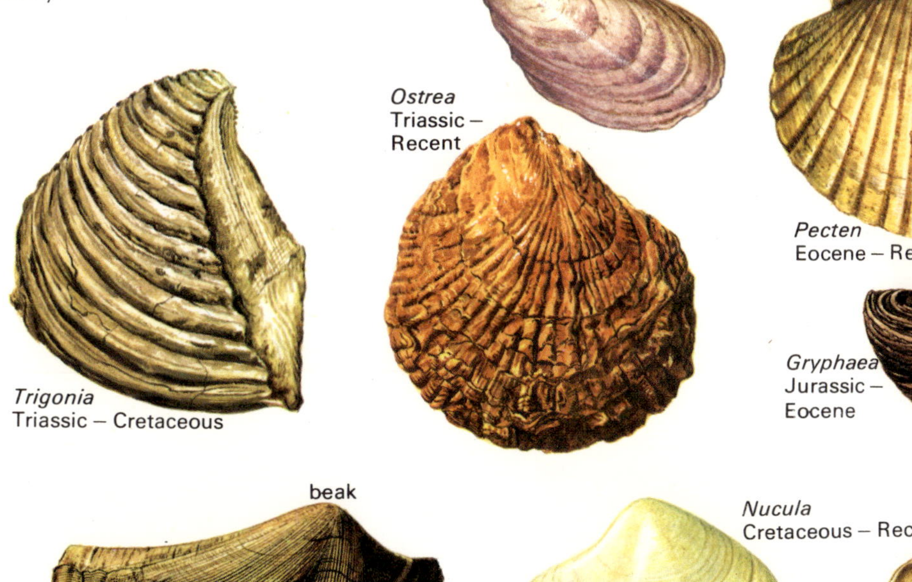

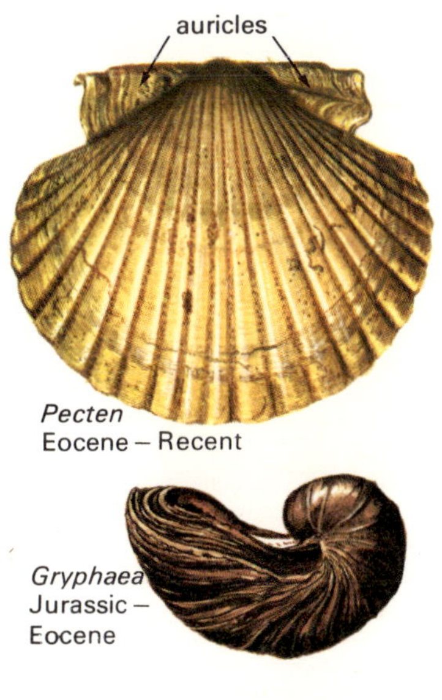

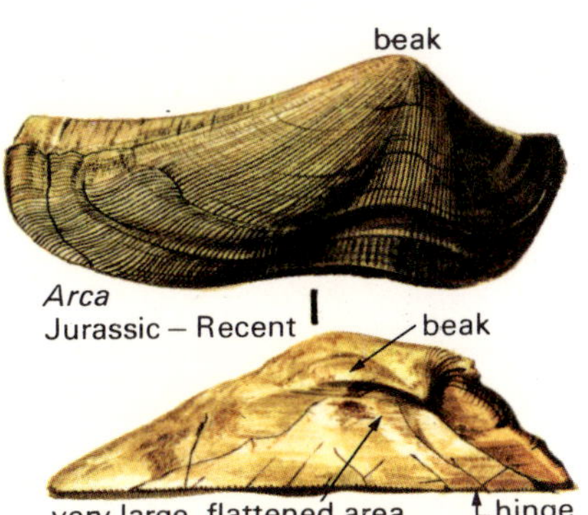

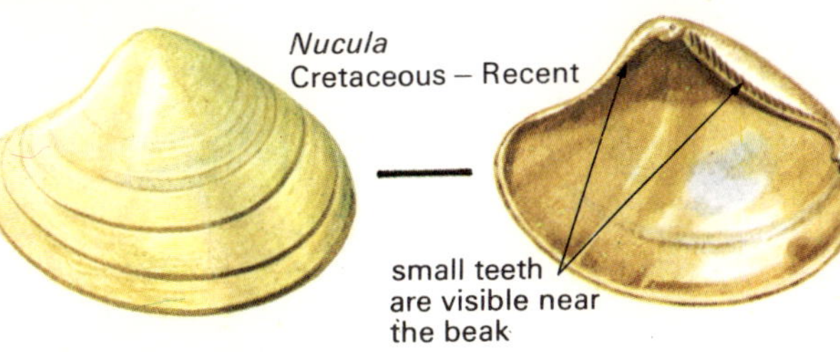

Scaphites
Cretaceous

Placenticeras
Cretaceous

Hamites
Cretaceous

Above:
Some typical fossilized cephalopods.

Right:
The structure of an ammonite shell.

Which group of animals are head footed?

This does seem a very peculiar question. In fact, the group is given the name Cephalopoda, which means head footed. The cephalopods also belong to the much larger group of invertebrate animals (do you remember the word invertebrate?) called the Mollusca or molluscs as they are commonly called.

The first cephalopods appeared in the seas of the Earth in Upper Cambrian times, and there are still members of the group alive to this day, although the modern cephalopods represent only a very small part of the group. Today's cephalopods include the squids and octopuses with their internal skeletons (contained inside their bodies) and the paper argonaut and the pearly nautilus with hard shells. You may have seen the beautiful shell of the nautilus when it has been polished and used as a decoration. Unfortunately, the persistent collection of this link with the past has meant that it has become very rare indeed. All of the members of the Cephalopoda, including those which no longer exist, live in the sea.

Perhaps you have seen an octopus in an aquarium. Even if you have not, you have probably seen pictures of them with their eight tentacles surrounding the beak-like mouth (hence the name, head footed). Many a sailor's story has told of giant octopuses dragging men or even whole ships to a watery grave, but most of them are quite small and harmless to man.

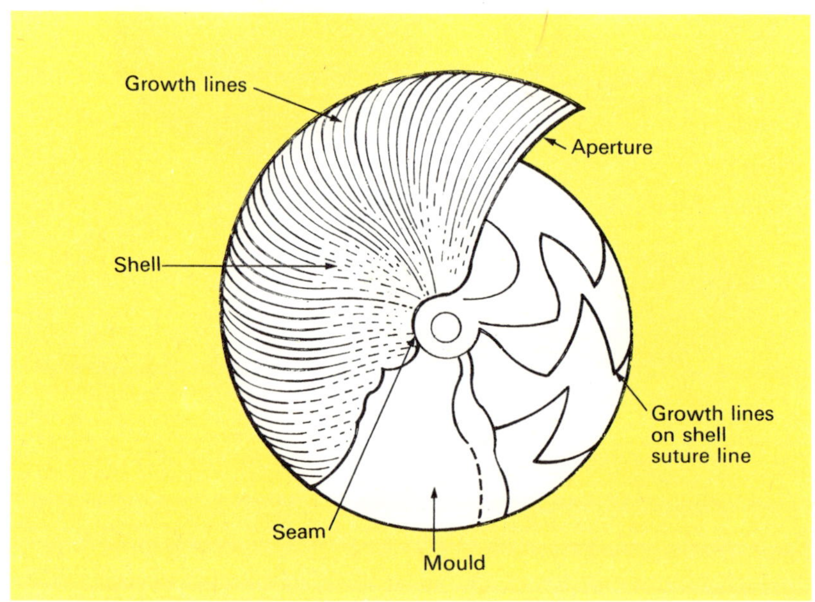

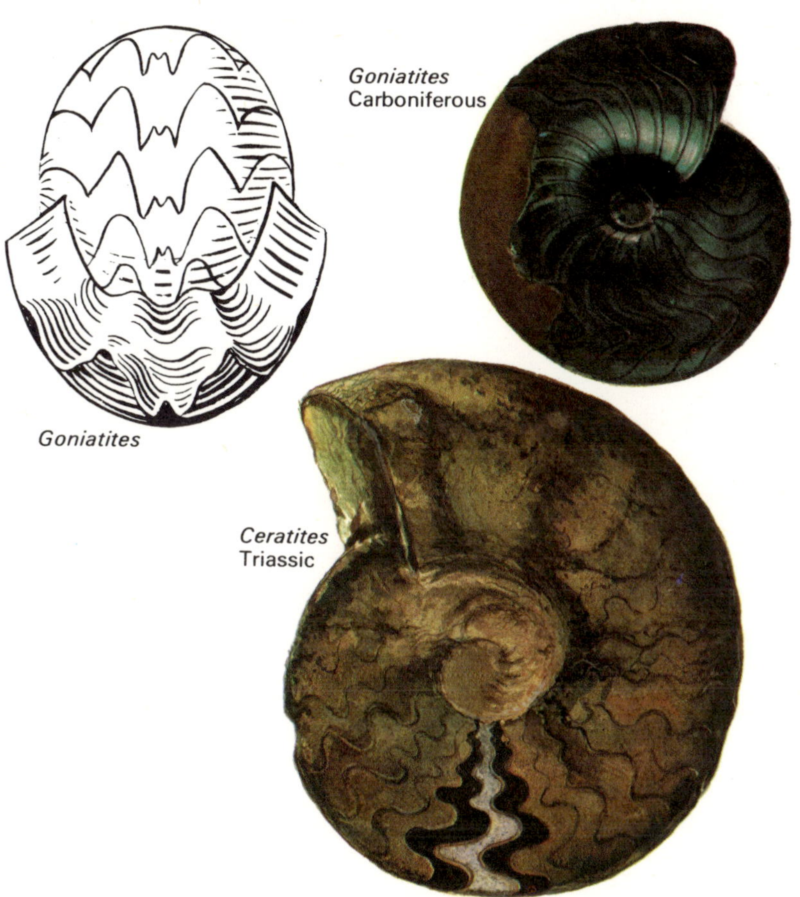

Right:
The markings on the shell of a cephalopod can reflect the internal structure.

The bodies of these animals are supported on a skeleton which is called a *guard*. It is said that cephalopods are among the most intelligent of the animals without backbones. It is certainly true that they change colour if a tasty morsel of food is put in front of them, so that they do seem to be able to show their feelings.

The members of the group that most interest geologists are the belemnites and, in particular, the ammonites. The belemnites had an internal guard which looked rather like a rifle bullet. These fossils are not particularly useful to those geologists who seek to label the ages of rocks with different species of animals. The ammonites are very useful, however, because they are widespread and common, and they developed different forms quite quickly. If you go to areas where there are rocks of the correct age, such as Lyme Regis in Dorset, England, you will be able to see many examples of these fascinating fossils looking like flat coils of shell lying on the beach.

What is coral?

If you have been on holiday by the sea, or if you live in a seaside resort, you will probably have visited some of the gift shops which are often to be found there. Apart from the postcards, pottery, and so on, you can generally find beautiful varnished shells of sea-dwelling animals, and the delicate pink and white fronds of china-like material which will be labelled coral. If you have ever visited places such as the north-east coast of Australia, you may even have seen a coral reef. But this does not answer the question. In fact, these pieces of coral and even the giant reefs are made by quite small animals which belong to a group called the Coelenterata. The hard coral is made of calcite which is secreted by the animal to form the platform on which it lives.

Corals are very primitive animals. The first evidence of their existence can be found in rocks older than the Cambrian which we have mentioned before. And, of course, you know from experience that they still exist today. The modern corals, though, are not exactly the same as the very early ones, as scientists have been able to prove from a close examination of fossils.

If you have looked in rock pools at the seaside, you will probably have had the chance to look at a relative of the corals in action – the sea anemones with their rings of

Right:
The structure of coral.

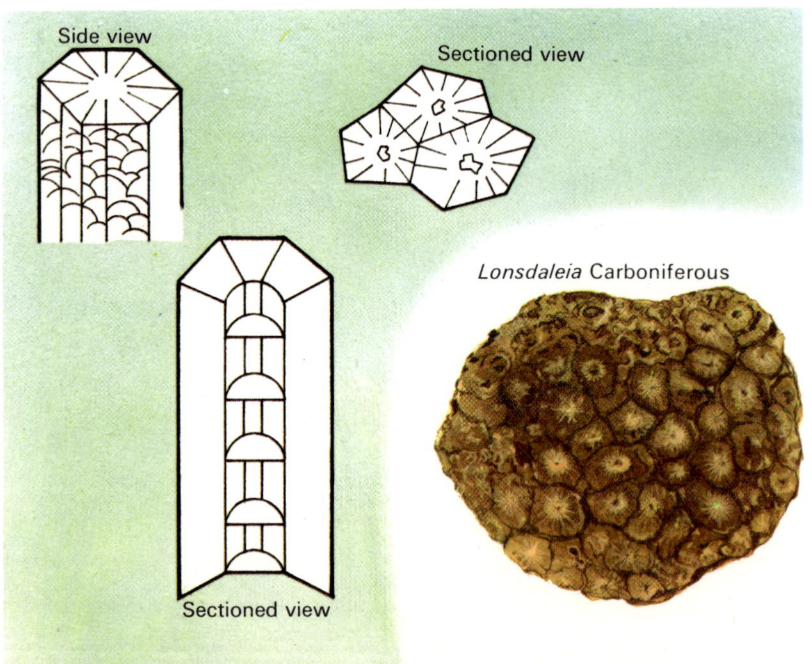

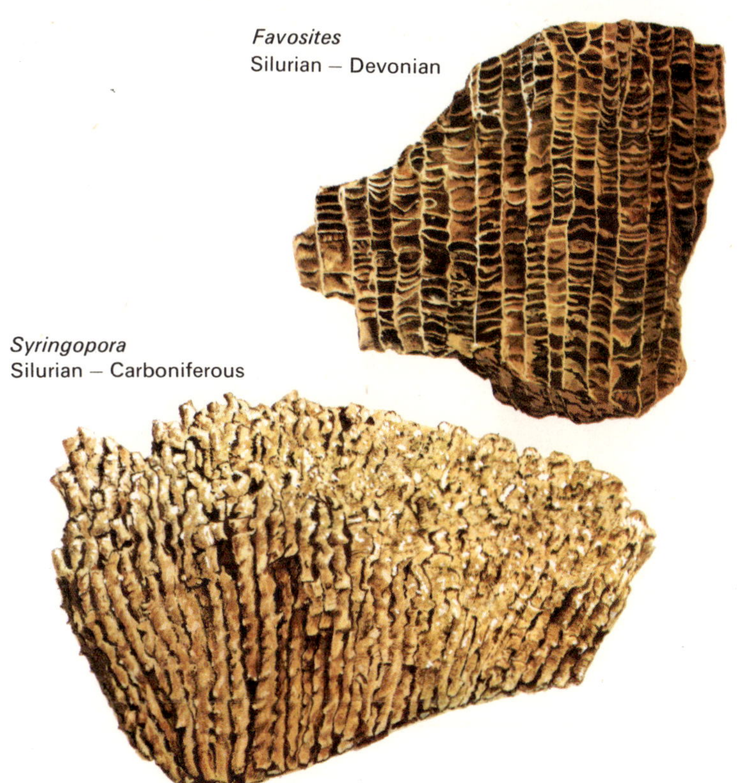

Right:
Two well-known fossil corals.

tentacles waving in the currents. If you imagine what an anemone looks like, you will have a good idea of the structure of a coral. The anemone is really very simple, consisting of its ring of tentacles surrounding the mouth which passes into the gullet and then into the main body of the animal which is made up of sheets of body tissue and muscles. As you know the sea anemone attaches itself to rocks and has no kind of skeleton. It is the hard supporting platform which the coral makes which can be preserved and is of interest to the geologist.

Corals breathe by absorbing oxygen dissolved in the water through the whole of the body surface, and feed by paralysing their prey with their stinging tentacles. They draw the victim into the mouth releasing any waste straight back through the same opening. Corals and sea anemones can move their bodies but generally stay in one place, but their relatives the jelly fish, can 'swim' by 'jet propulsion', forcing out a jet of water.

There were many different species of corals throughout the fossil record; some types which lived in colonies and others which lived alone. Fossil corals are found in limestones. Coral limestones are often cut and polished to make very attractive building stones. Limestones found near Torquay, in Devon, England are famous coral limestones.

Which are the most advanced animals without backbones?

Holocystites Silurian

Above:
A fossil echinoderm called a cystoid.

What do we mean by advanced? In this case we mean those invertebrates which are most like animals with backbones including man himself. This does not necessarily mean that all the other groups of animals without backbones are not very efficient living creatures, capable of great variety and different ways of life.

The group of sea-dwelling invertebrates called the Echinodermata are the most advanced in our sense of the word. Echinodermata means spiny skinned. The word refers to the spines which protect these animals and on which they move.

There are two main divisions of the Echinodermata. Both these divisions have modern, living representatives. There are the free-living forms (meaning that they are not doomed to stay in one place) which include the starfish, sea cucumbers, and the sea urchins. There are also the forms which are attached in some way to the sea floor, with names like crinoids and blastoids. All the members of the Echinodermata have skeletons of calcite buried within the soft tissue, although sometimes these skeletons are rather reduced in size.

Many invertebrate shells are used for protection. Because the skeletons of this group are internal, they act as a support for the body like our own framework of bones. This internal skeleton can become fossilized and is of interest to geologists.

The two groups which are of most importance in the fossil record are the sea urchins or echinoids, and the crinoids or sea lilies. It is worth noting here that sea lilies are the common name given to these crinoids. It is an attempt to describe

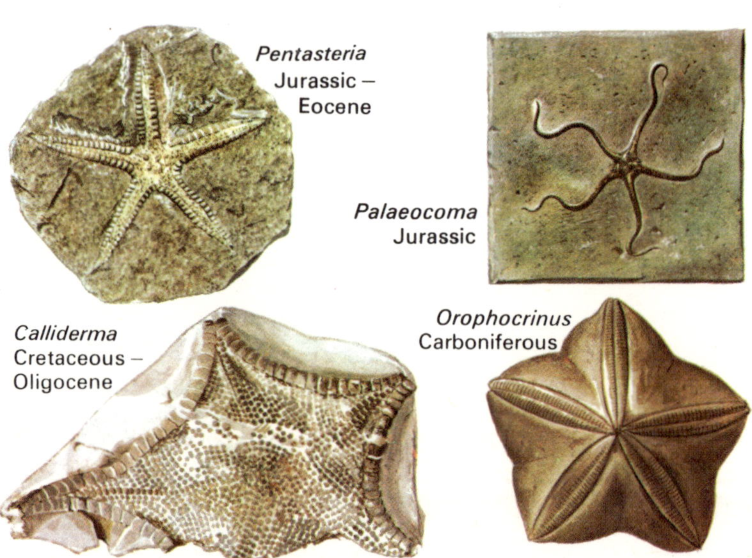

Right:
Some fossil starfish.

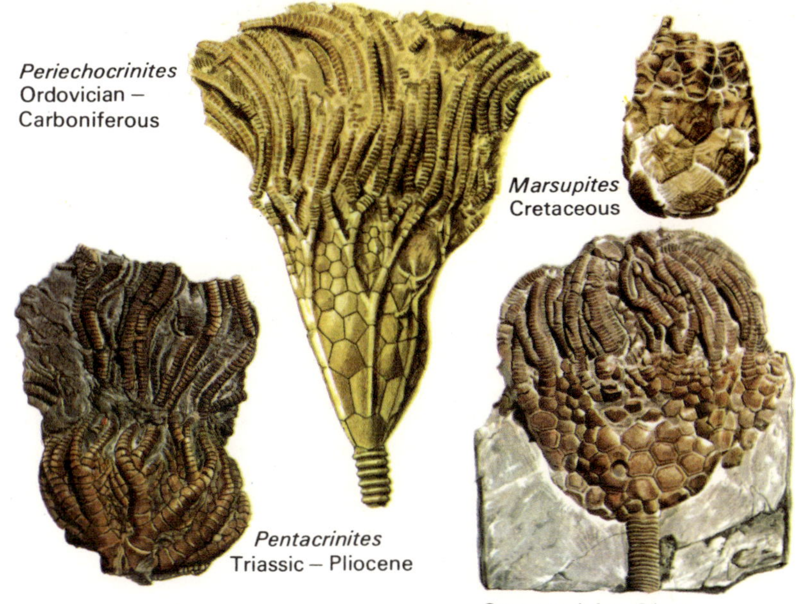

Left:
Fossil sea lilies or crinoids.

their appearance, and you must remember that they are animals, not plants.

The skeletons of sea lilies are made up of a large number of separate units called ossicles. When the animals die these ossicles often become separated so you would be quite fortunate to find a whole fossil crinoid. You are more likely to find the individual parts, but even these tiny fragments can be positively identified.

The skeleton or test of the echinoid is made of plates of calcite and the animals grow by adding more calcite to each plate. Spines also develop by accumulation of calcite. This skeleton is very strong because of the way in which the plates are arranged, and whole fossils may be found, for example, in flints in the chalk. Echinoids first appeared about 500 million years ago during a period which we refer to as the Ordovician, and there are still sea urchins in modern oceans.

Below:
The fossil on the far right is a fossil echinoid or sea urchin and it is known as an irregular type. These are much more common in the fossil record than the two regular types also illustrated.

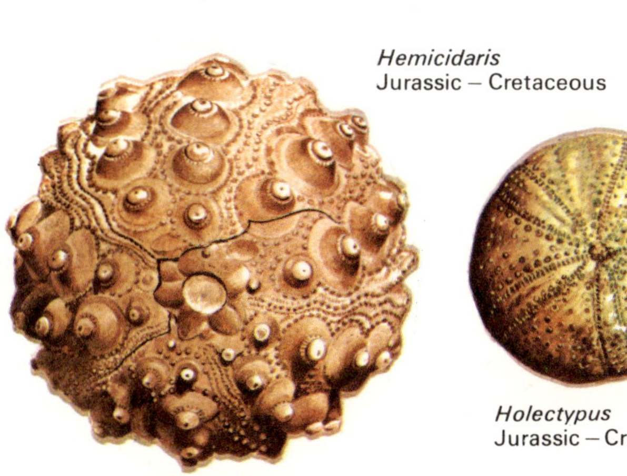

How have animals with backbones evolved?

You have seen that the animal kingdom is divided up into a number of major groups. Each group is usually referred to as a *phylum*. Man belongs to the phylum Chordata. All animals belonging to this group must possess a *notochord*, that is, a structure like a spine at some stage during their life cycle. In some animals, such as the graptolites that have already been discussed, this notochord remains in its elementary state. In the case of a group of chordates known as the Craniata, this notochord develops into a support for the body, and the skull or cranium which protects the central nervous system. As you might have guessed by now, the Craniata includes all the birds, mammals, and so on, and of course, man himself.

The Chordata are thought to be very closely related to the Echinodermata; the group of animals without backbones which includes sea urchins and starfish. This can be proved by the great similarity of certain chemicals contained in the bodies of both groups. It is certain that all of the ancestors of the chordates lived in the sea.

The earliest vertebrates – animals with backbones – were probably living as long ago as 470 million years. Their

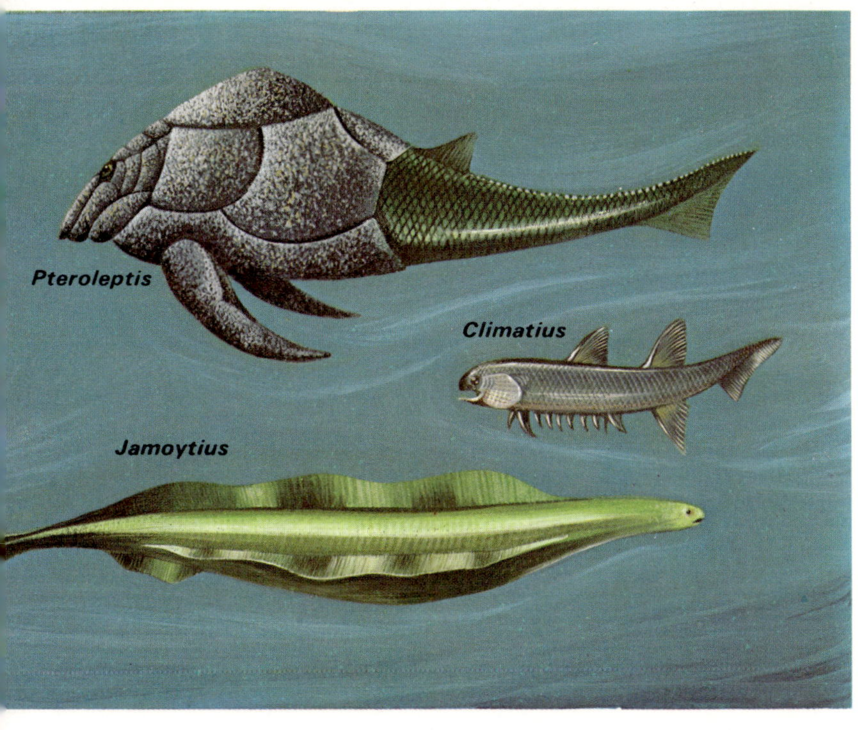

Left:
An artist's impression of some of the earliest known vertebrates.

remains can be found in North America in a bed of rock known as the Colorado Bone Bed – so called for obvious reasons! This rock is a marine sandstone, and other fossils can be found in the same rocks, including occasional brachiopods, and still more rarely, trilobites. The bones are very worn which indicates that the conditions were quite turbulent, such as you might expect to find along a shoreline today.

These earliest vertebrate remains are the bones of primitive fish, and it is thought that they lived in the place where they are found. It is interesting to compare these remains with the bones of modern jawless fish. Modern jawless fish do not have effective kidneys because the salts in their bodies are in balance with their surroundings. It is reasonable to suppose that the same was true for their early vertebrate ancestors. This sounds even more likely when we take into account that there is more energy in the form of food available in the sea so that evolution is more likely to come about.

The first backbone simply acted as a support for the fish's muscles, but later it evolved to protect the brain. The first fish were armoured fish. The backbones should not really be called bones at all, because they are made of cartilage, and bone merely provided the fish with protective armour. From that time onwards vertebrates evolved very speedily indeed.

What animals developed from the armoured fish?

You have seen the way in which the early fish had a spine made of cartilage, and that bony plates served to protect them. They probably needed to be armoured because they could not swim particularly well. As they became more mobile, however, the need for the protective armour was reduced. The bone cells slowly began to move inwards, changing the cartilage into bone. Thus, the first fish with a true backbone evolved. It is interesting to note that modern sharks still have a spinal column which is made of cartilage because the ossification, as the change to bone is known, has not occurred in this group.

The first bony fish appeared during early Devonian times, that is, about 380 million years ago. At the end of Devonian times, about 350 million years ago, the bony fish began to evolve very rapidly indeed, and divided into two groups. The groups are the modern bony fish, which account for about 95 per cent of all modern fish, and the remaining five per cent are a group known as lung fish.

The modern bony fish form a group in which there is just one single line of evolution. This evolution takes place as a gradual loss of the heavily armoured scales together with a change in the structure of the jaw. The change in the jaw structure is mainly concerned with the way in which the mouth opens, but it is rather complicated and it is enough to say that modern fish are able to open their mouths far wider than their early ancestors. This ability has obvious advantages when it comes to feeding.

There are still lung fish living today. They differ from more familiar fish in two main respects. Firstly, they have internal nostrils rather than gills, and equally important, the structure of the fins is different.

Right:
An early amphibian.

Below right:
A lobe-finned fish.

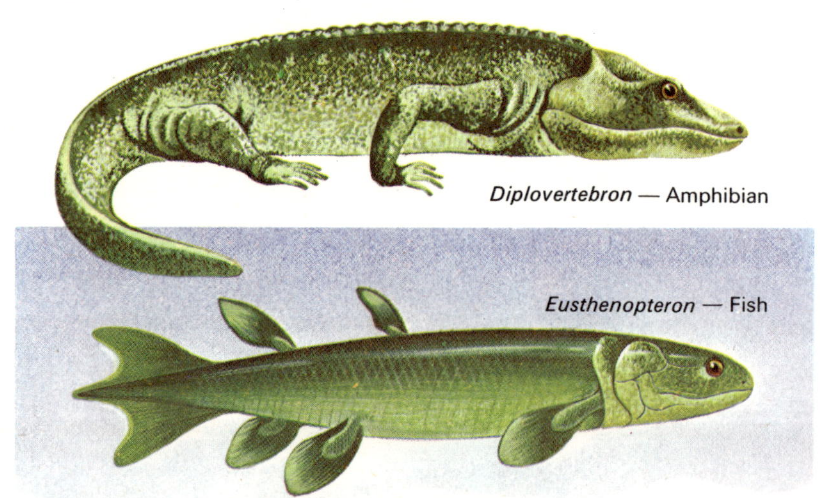

Right:
A specimen of a living coelacanth was found off the South African coast in 1939. This, and a number of other examples that have been found since, show that this fish has existed for as much as seventy million years.

In the true fish, the fins are formed on rays which extend from the body – they are usually referred to as ray finned. In the case of the lung fish, however, the fin rays enter an extension of the body tissue itself – they are known as lobe finned. This means that the fins of the lung fish eventually led to their ability to move on to land because of their development into legs.

Modern lung fish are able to survive for long periods when ponds dry out. They burrow into the mud at the bottom of ponds, only to emerge when the pond fills with water once again.

During Devonian times, the widespread desert climate tended to encourage the development of animals which were able to survive on land as well as in water. These animals were the amphibians. They first appeared in Greenland, which was furthest away from the return of the sea during Carboniferous times some 300 to 350 million years ago.

Right:
Just a few of the primitive fishes that were living during Devonian times.

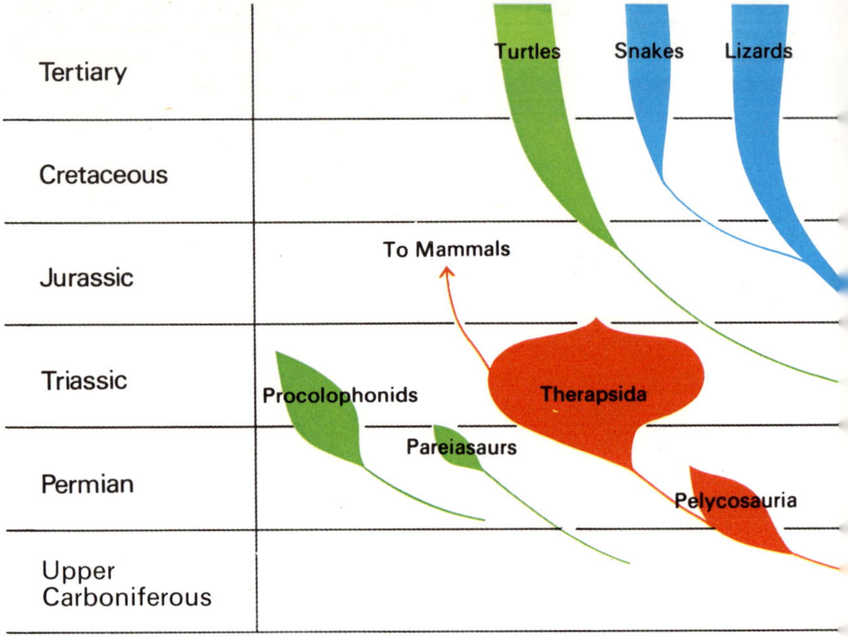

Tertiary
Cretaceous
Jurassic
Triassic
Permian
Upper Carboniferous
Turtles
Snakes
Lizards
To Mammals
Therapsida
Procolophonids
Pareiasaurs
Pelycosauria

After the amphibians – what next?

The amphibians were able to survive on land as well as in water, which as we have seen, was of considerable advantage during the arid climates of Devonian times. Nevertheless, all amphibians, even those that survive today, must return to water to breed. In fact, after the amphibians had developed, and seemed to have come to terms with their new way of life, there was a mass return to the sea, and few amphibians still exist. The early amphibians were large, being as much as two metres long, but gradually they became smaller and more variable in appearance.

The amphibians that returned to the sea had limbs and not fins like the fishes. The body became much longer and most species of the newly returned marine forms became predatory, that is, they hunted other animals for food. A new group of animals evolved in the late Carboniferous times, some members of which were to dominate the Earth for a considerable period of time. This new group is the reptiles. Land forms perfected their means of moving in their dry surroundings by increasing the strength of the limbs.

By the end of Carboniferous times drier conditions once more came to the Earth. Thus, those species which were better able to cope with these conditions survived while many became extinct. The reptiles developed an extra outside layer to prevent the body from drying out, and eventually a means of protecting the delicate egg was found. Reptiles' bodies became covered with hard plates, and these

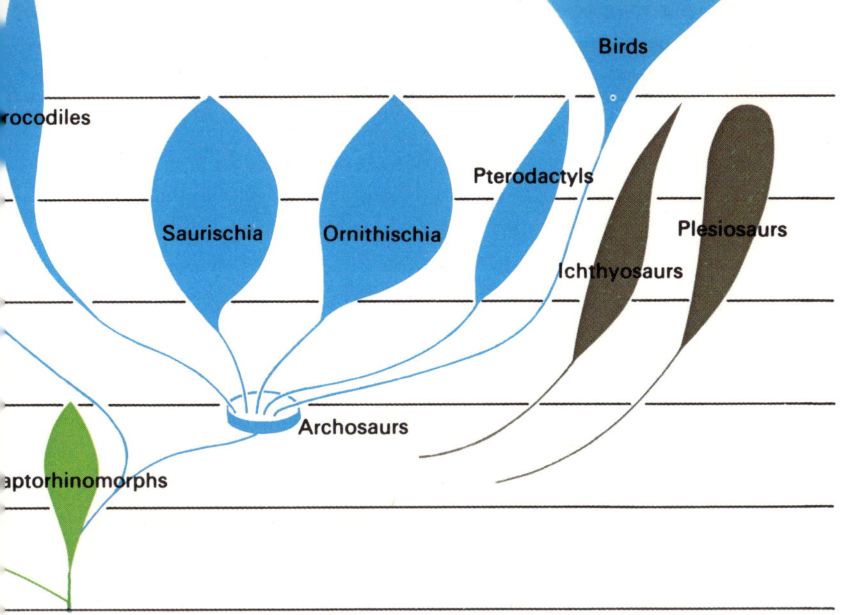

Left:
This diagram shows how the various groups of reptiles (some of which have very complicated names that need not concern us here) are related to one another throughout the fossil record. It also shows how the birds evolved from reptiles.

are well shown in today's relics of the past, the crocodiles and alligators, for example. It is possible that the protected egg, capable of withstanding dry conditions, came first as a protective adaptation for one group of amphibians, but this is not certain and there is much speculation about these early amphibians and reptiles.

During Permian times where dry conditions were very widespread and deserts were far greater in number and extent than they are today there was a very rapid evolution, in which all the various habitats were occupied. It is worth a mention here that at the same time a group of warm-blooded animals were also beginning to evolve – the mammals. It would be some time, however, before they would be as powerful as they are today. There were also mammal-like reptiles, and in some areas, such as South Africa, these were the dominant species. Instead of the flat scales of the reptiles, the mammal-like reptiles had fine scales standing at an angle from the body and which would eventually become hair. Hair acts as a better insulation against heat loss, so that they could live in cooler climates and eventually develop a warm blood stream so that there would be no need to hibernate during winter. But for 100 million years the reptiles were dominant, and one group, the dinosaurs, were the unchallenged masters of the Earth.

Left:
Hylonomus, one of the earliest known reptiles.

Which animals are among the largest that have ever lived?

It is probably safe to say that some of the largest animals that have ever lived on the surface of this planet are members of a group of animals which have been collectively called dinosaurs. These 'terrible reptiles', as their name means,

have excited the imagination of men ever since bones of the first dinosaurs were discovered in rocks of Upper Triassic age – that is, deposits about 200 million years old. Many tales have been told of these monsters sharing the Earth with man, but, of course, as we shall explain later all the dinosaurs had died out long before man appeared.

Dinosaurs evolved into a great variety of different shapes and sizes but they all had a common ancestor. They ranged in size from dinosaurs the size of a chicken, to others which were as much as 30 metres long and weighing more than 30 tonnes. Some dinosaurs, such as *Diplodocus*, lived on plants. Others were hunters. They attacked and killed dinosaurs as much as 15 metres long for food. *Tyrannosaurus* were the largest meat-eating animals that have ever dwelled on land. Bones of these animals have been discovered in rocks about 70 million years old in North America. There were also dinosaurs that lived in the sea. Others such as *pteranodon* developed a kind of wing and they could glide in the upward currents of air that occur above the cliffs on which they must have lived.

Some of the less terrible dinosaurs could escape the predatory ones using their speed, but others were much too clumsy and slow-moving to get away. They developed armoured coverings to their bodies to protect them and had spines and horny plates as well. In fact, *Stegosaurus* was probably able to

defend itself quite well with the help of spikes at the end of its long tail.

Dinosaurs have presented researchers with many puzzles, not least of which is how giants like *Brontosaurus* managed to live at all. Examination of its bones has shown that its legs were probably not strong enough to support its weight on land, and if it lived partly submerged in water, its rib cage may have collapsed. On the other hand, it did not possess the correct kind of feet to survive in swamps.

There have been many arguments about why the dinosaurs suddenly died out about 65 million years ago. It has been suggested, for example, that an increase in cosmic rays led to them laying thicker and thicker shelled eggs, so that eventually the young could not hatch.

Tyrannosaurus

Which were the first animals with warm blood?

The ability to control the body temperature is a considerable advantage to an animal. Even today, reptiles are cold blooded. In practice, this means that their body temperature corresponds with the temperature of the surrounding environment. Now, the efficiency of an animal's body functions is very dependent on its temperature. This means that as the air temperature falls, the animal's body temperature falls and its functions become more and more sluggish. Thus,

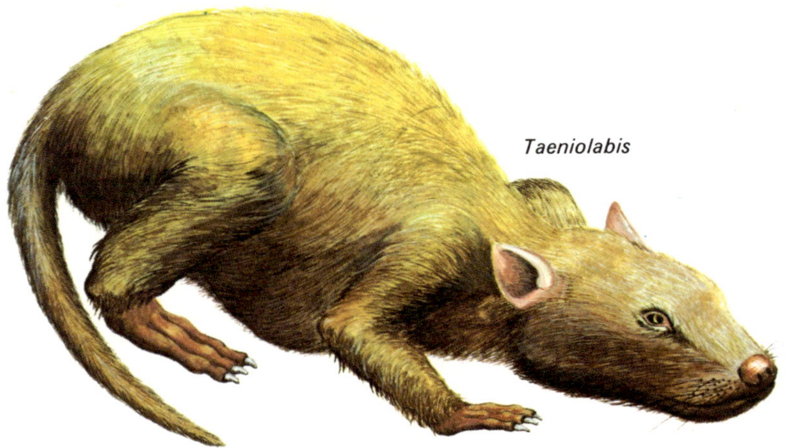
Taeniolabis

reptiles survive perfectly well in warm climates, and even in cooler climates they manage by hibernating during the winter months. But cold blooded land animals cannot survive in the very cold regions of the Earth. You can see, then, that if an animal has the ability to control its body temperature it has a considerable advantage over the reptiles in its adaptability to cold climates. It does mean, however, that in cooler places, animals must have their bodies insulated to prevent loss of heat. This can be achieved by extra layers of fat, hair or fur in the case of mammals, and feathers in the case of birds.

The ability to control body temperature may have evolved separately on two occasions. It is possible that a group of primitive reptiles may have been partly warm blooded and probably the dinosaurs could control their body temperature to some extent.

An animal with warm blood has a very high metabolic rate, as the speed of its body processes is known. But this in turn means that their cells, which make up the bodies of all living things, require more oxygen or they may be damaged. Together with warm bloodedness, then, the blood supply carrying oxygen around the body must be made more efficient. There must be more blood vessels in the bones. It is interesting to note that the bones of dinosaurs have almost as many pores as those of modern mammals.

In late Jurassic times (about 150 million years ago), an animal existed which has great importance in the development of life on Earth. This creature is known as *Archaeopteryx*; we expect that you have heard of it. It seems to be exactly intermediate between a bird and a reptile. It had some feathers and could probably fly, although not very well. In order to be efficient enough to fly it must have been warm blooded. But its skeleton is more like that of certain dinosaurs, and rather than a bill, its jaws were armed with small teeth. This first bird probably developed from some of the reptiles that leapt from tree to tree.

Morganucodon

Ctenacodon

Pantothere

What makes a mammal a mammal, and how do they develop?

Think of a typical reptile, for example a lizard, and a typical mammal, a fox. From your own experience, in what ways do you think that they differ? Firstly, a lizard has a tough, leathery skin, whereas the fox is covered with fur. You have learned already that reptiles are cold blooded, but mammals control their own body temperature, so that the fox needs its fur to prevent heat loss during the cold parts of the year. If you go out on a cold day without a top coat, you probably shiver. This rapid shaking of the muscles is a mammal's way of warming up the body should it become too cold. On the other hand, if you do any strenuous exercise on a warm day, you will sweat. The moisture produced evaporates from the skin causing the body to cool. A reptile must rely on the surrounding air temperature to keep up its body heat.

Of course, a fox has cubs; in other words the 'egg' of the fox develops inside the vixen (as a female fox is called), and she then gives birth to living young instead of an egg. The vixen must then lavish a great deal of care in bringing up her offspring, firstly feeding them with her own milk and then teaching them how to fend for themselves. The lizard, however, lays eggs, and when the young hatch they are very active and able to look after themselves quite well. Here, then, we see another important difference between mammals and reptiles.

Perhaps the most important characteristic of mammals is their great intelligence. Remember that man is a mammal, and it is his intelligence that has enabled him to develop to such an extent and have such far-reaching effects on the world in which he lives. Compared with a dinosaur such as *Brontosaurus* man is obviously very small, but in relation, his brain is very large indeed.

Most of the actions of reptiles are conditioned by the need to feed, sleep, and reproduce, and these actions do not change markedly throughout the life of an individual. A mammal, on the other hand, can commit things to memory; it can learn from its experiences, and its future behaviour will

Left and far right: Modern reptiles include lizards, some examples of which are shown here.

Right:
The giant panda is a mammal. Mammals are able to control their body temperature and give birth to live young which the mother rears.

be altered accordingly. You know how easy it is to train a dog (if the training method is correct) to behave in the way you wish rather than in the way that would be natural for a pack animal.

There are many other things that make up a mammal but, if you like, almost all of them are improvements on the simplest type of vertebrate. In other words, the modifications that have taken place have been to make the animal more efficient, more able to feed and protect itself, and to adapt better to its environment or even change the environment to suit itself.

How have mammals evolved?

As you know the group of animals known as mammals are successful, widespread, and very diverse. Modern mammals range in size from animals such as the lesser shrew of Britain at a total length of a little over six centimetres and weighing little more than 4.5 grams to the giants of the oceans such as the blue whale weighing in at up to 150 tonnes.

Modern mammals occupy a range of different habitats: they can live in deserts, as does the desert fox, in frozen wastes as does the polar bear, in jungles where many species abound, in temperate woodlands like the European badger and so on. These are the mammals that live on land. As we have seen, there are those that roam the oceans such as the whales.

Right:
The whale is a mammal that has returned to a marine way of life.

Ocean dwelling mammals have more problems than their counterparts on land because they still have to maintain their body temperature. Obviously hair would be useless to help them retain heat. Consequently, they have evolved thick layers of fat usually known as blubber. On the other hand, because the salt water helps them to support their body weight, they have been able to grow to these enormous sizes. In fact, the largest whales are the largest animals that have ever lived, larger even than the mighty, vegetarian dinosaur, *Brontosaurus*.

Mammals have also conquered the air, perhaps not as efficiently as birds, but nevertheless bats could still be considered to be successful animals in the struggle for survival.

How, then, has this great diversity been achieved? In Triassic times, some 200 million years or so ago, mammal-like reptiles had developed at least partial control of their body temperature and may have been covered in fur or hair rather than scales. It is difficult to say with certainty from fossil evidence, exactly when the first true mammals

The duck-billed platypus (*above*) and the spiny ant-eater (*right*) are the only surviving mammals that lay eggs and they are very primitive indeed.

appeared, but it was certainly not for another thirty or forty million years, when at the end of Triassic times animals, not unlike modern shrews, were alive.

The most primitive mammals probably still laid eggs rather than giving birth to living young, and today there are still mammals surviving which are also egg-laying. The duck-billed platypus of Australia, for example, lays its eggs at the bottom of a burrow. It has been able to survive (it could be described as a living fossil) because the Australian continent has been cut off by the drifting apart of the continents for at least 100 million years. Marsupials, also found mainly in Australia, represent a kind of intermediate stage of evolution with the young being born alive at a very early stage and remaining in the mother's pouch.

From Tertiary times until today, in geologic terms, the evolution of mammals has been very rapid with the development of animals such as the horse and the elephant being the easiest to trace.

Tasmanian wolf

Kangaroo

Right:
The Tasmanian wolf and the kangaroo are examples of marsupials that have survived in Australasia because this area became separated from the rest of the world by continental drifting. As a result, these animals had little competition from more advanced mammals that were developing elsewhere.

When did plants first colonize the land?

We know already that the first life on Earth was probably plant and that it lived in water. Indeed, these are the very primitive algae that we have already mentioned. You must bear in mind that these first plants did not have stems, leaves, and roots as do the plants that you might immediately think of as living today. These plants were just collections of living cells.

In rocks of Cambrian age, however, there is evidence of plants which might have lived on land. This was between 500 and 600 million years ago. But it was not until about 150 to 200 million years later that plant life became abundant on land and evidence of this can be found in rocks of the so-called Silurian and Devonian periods.

The first plants that began to encroach the land had to live in shallow water, but they did have roots and could take in the gas carbon dioxide from the primitive atmosphere. But the first true land plants were a group which are called *psilophytes*. They still had to live in damp places, however, and they do not seem to be directly related to any plants living today.

Above:
An artist's impression of a plant that has been found fossilized in Devonian rocks of Scotland. It lacked the true roots, leaves and seeds of modern plants.

Right:
The modern club moss and the fossilized fragment of a lepidodendron shown here belong to the same group of primitive plants.

The first trees appeared in the Devonian, although they were not much like the trees that you will be familiar with in a modern country landscape. For one thing these trees, with the group name of *lepidodendrons*, did not produce flowers, and in fact it was not until a mere 135 million years ago that the first flowering plants became abundant. Modern examples of very primitive land plants are the ferns and mosses. About 350 million years ago, however, land plants became so abundant that their remains have left us with the deposits of coal which are so important to our industries and help to warm many of our homes.

Right:
These horsetails, ferns, seed ferns, and tree ferns can be found fossilized in rocks of Carboniferous age.

What is the importance of plants in life's framework?

It is only when we stop to think that we realize the importance of land plants to us. We eat them in our cereals and vegetables, we make use of them for fibres like cotton, we depend on them for many drugs such as penicillin, and as we have already mentioned we need the fossil fuels. But even when we eat meat, we must remember that the animals which produce the meat probably lived on plants.

Plants are nature's producers. From sunlight, oxygen, and other foods from the soil, plants manufacture the foods upon which all animals ultimately depend. It is worth noting that it was with the rise of the land plants in the Devonian, that land animals also began to evolve.

What is known about ancient plants?

We have looked at the ways in which plants first colonized the land surface of the Earth, but perhaps it is time to look at the evolution of plants as a whole in more detail. It is generally easier to follow the trends in the evolution of plants because modern advanced plants may be quite similar in their make up to a more primitive form. Animals, and the development of the animal kingdom are very much more complicated.

As has already been mentioned, plants are made up of *cells*. These cells are the living substance from which all life is composed. It is unnecessary here to discuss the detailed organization of living cells. It is enough to say that they are made up of a dense nucleus surrounded by a watery fluid. The very first plants, of course, were probably just single cells of living matter.

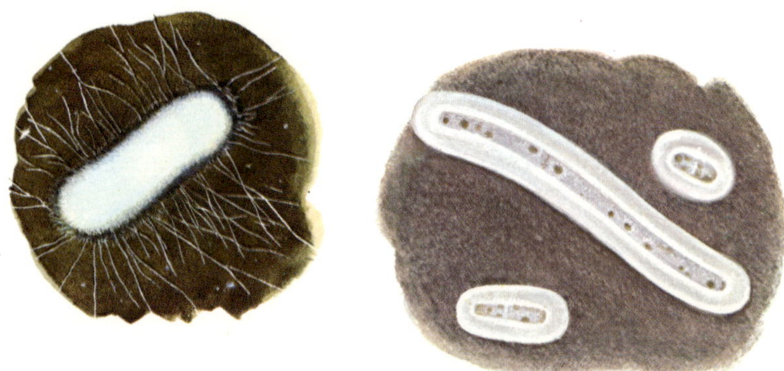

Right:
Two true bacteria. It has been stated that bacteria reproduce so quickly that if an unlimited food supply was available the weight of bacteria produced after twenty-four hours from a single bacterium would be greater than the mass of the Earth.

Bacteria are sometimes thought of as plants. These are tiny (microscopic) single-celled organisms which usually live by feeding on other plants or animals as parasites, or on dead or decaying matter. It has been discovered however, that there are bacteria that will even feed on such unpalatable things as concrete and glass. In fact, without bacteria, no other life would be possible, because it is bacteria that aid in the breaking down of rock debris to form soil, and bacteria that enable waste products to decay. As you might expect, fossil remains of bacteria are rare, although they can be found in the rock called the Rhynie chert of the Scottish Devonian.

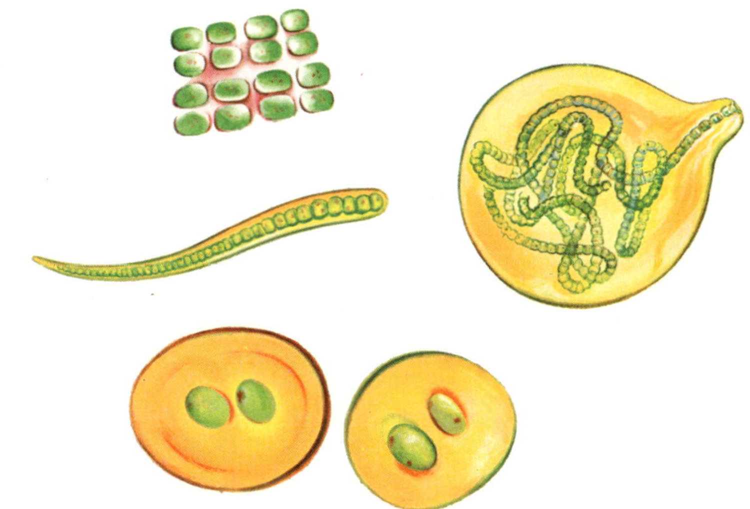

Right:
It is thought that these blue-green algae are important in the formation of soils.

The algae are also a very primitive group of plants, still in existence today, but representing some of the earliest life forms to be found on Earth. The algae are the most diverse plant group both in size and in their ways of life. Remember that seaweeds are algae as is the green film that forms on the glass of a goldfish tank.

The evolution of plants can usually be traced on the basis of their way of multiplying, and the algae are no exception. Their method is very primitive. At a particular time of the year the cell contents become more dense, two cells come together, and new nuclei develop. The rest of the year the cells just divide.

Algae are usually subdivided into groups on the basis of their colour. Green algae are those that can only live in clear, comparatively shallow, marine conditions and they build reefs. Red algae include the seaweeds and live in the deep seas making use of only part of the light that is found at depth. The blue-green algae are thought to be important in the formation of soils. The brown algae live in shallow water marine conditions.

What happened to plants next?

So far we have considered the very first primitive plants that existed on Earth many hundreds of millions of years ago, and we have seen when they first appeared on land and the effects that this had. Now perhaps, we should look at the more advanced forms including most of those plants that we are so familiar with.

The group of ancient plants that forms a link between advanced and primitive plants is known as the *Cordaitales* which first appear in the Lower Carboniferous rocks some 300 to 350 million years ago. This group reached the height of its development in terms of both numbers of individuals and species at the time when the last of the important coal deposits were being laid down, that is, in the Late Carboniferous times about 270 million years ago. They could be thought of as being intermediate in type between epidodendrons mentioned earlier and modern conifers. (Remember that conifers are trees that bear cones, such as larch or Scots pine.) All the members of the Cordaitales were tree-like, and were able to withstand very dry conditions. They had leaves that were not flattened so that less water was lost through them. The outer 'skin' of the leaf became much thickened, and the shape of the tree was such that air became trapped within the branches.

In Middle Permian times, about 250 million years ago, true conifers were becoming quite abundant, and almost 100 million years later in the Jurassic period they had reached their peak. It can be shown that conifers evolved directly from the Cordaitales. Evolution of the conifers occurred in two different directions. Members of the genus *Pinus* became

Below:
A section through a typical flower.

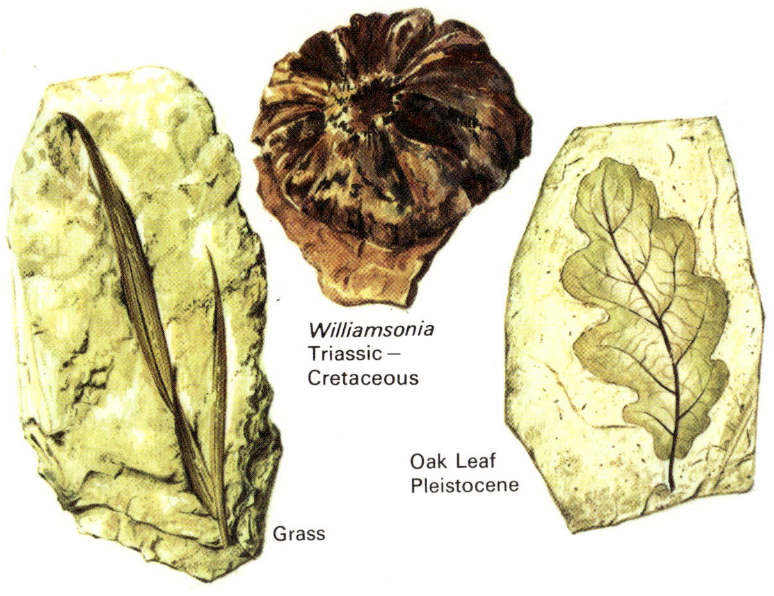

Left:
Here are some fossil remains of flowering plants. Remember that trees such as oaks and horsechestnuts are flowering plants as well.

Below:
Some common coniferous plants that can be found today.

Scots Pine

more and more able to withstand drying, indeed, would not survive in damp conditions.

Another group, known as the *Taxus-Macrocarpa* group, continued to retain the more primitive type of flattened leaf. Members of this group could not withstand drying but could survive in very poor soils where there were strong winds and high rainfall or snow – that is, mountain conditions. All conifers must live in association with a root fungus because they do not have root hairs.

There are two other groups of plants that seem to have developed from the early types, but this is not certain. They are both palm-like types, but one group, the *Bennetitales* became extinct during the Cretaceous period, about 100 million years ago.

The plants that you are probably most familiar with are the flowering plants. These probably evolved indirectly from the conifers at some point during the Jurassic, but they did not become important until the Tertiary, between 50 and 70 million years ago. They have not yet reached their maximum development.

Juniper

Monkey
Puzzle

Which animals are among the tiniest that have ever lived?

Among the tiniest animals that live independently today are members of the *plankton* which inhabit the seas everywhere. These *zooplankton* as they are more properly called are very important to the whole life of the sea because, together with the tiny plants upon which they feed, the *phytoplankton*, they provide the nourishment for all the other animals which live in the sea. Even giant whales feed upon plankton.

Right:
Some modern foraminiferids viewed with the aid of a microscope.

Right:
Radiolarians like the ones shown here are planktonic marine animals which may feed on tiny plants.

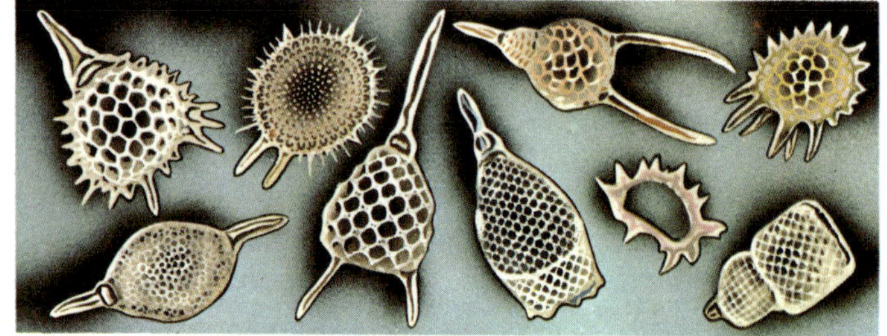

For the geologist, however, there are other groups of microscopic animals (this means that they can only be clearly seen with the aid of a microscope) that are more important. Firstly, there is a group of animals called *ostracods* which range in size from one millimetre to two centimetres so that they could be seen with the naked eye. Ostracods belong to the arthropods which include insects and spiders. Do you remember where you have seen this word before? Remember that insects and spiders are arthropods, but which fossil group do they belong to?

Right:
These are the parts of the radiolarians that may collect on the sea floor to form a radiolarian ooze. This only forms at great depths – perhaps greater than 4500 metres.

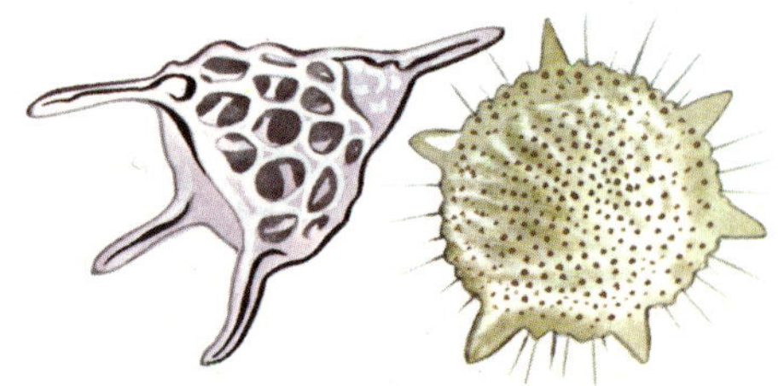

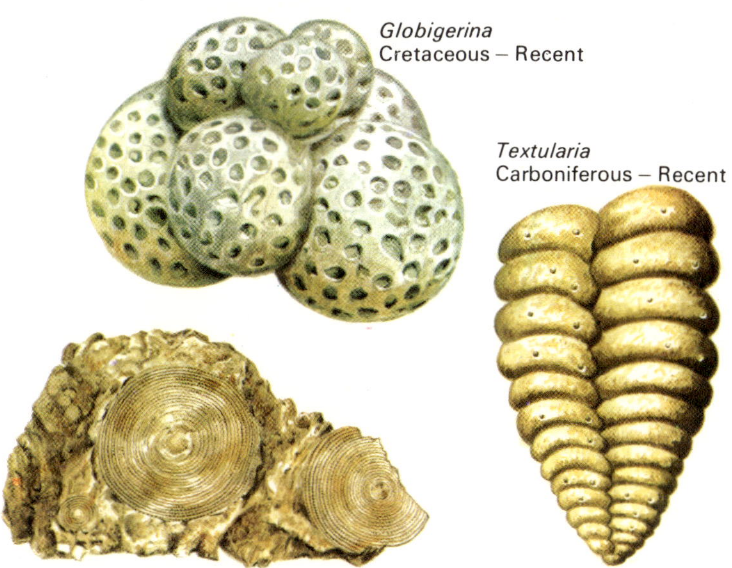

Right:
Some fossil 'forams'.

Ostracods are different from all the other arthropods in that they have an external hard shell in the form of two parts or valves rather like that of the bivalves that we have already mentioned. The actual animal living inside its shell is not unlike a modern lobster. As it grows, the ostracod sheds its shell in a series of moult stages.

Ostracods all live in water, but it may be in stagnant fresh water or in rivers or in the sea. In fact, a single species can tolerate quite a wide range of conditions. They mostly live at the bed of the sea, river or pond and like the trilobites they are scavengers. The first ostracods can be found in rocks that are 550 million years old and representatives of the group can still be found today.

Another group of microscopic animals that are of special interest to the geologist are the *foraminiferids*. They are members of the Protozoa, which all consist of only one living cell, and they are important because they produce a skeleton which can be fossilized. They are all marine animals, but different types may have lived on the sea floor and some may have been free-floating. The skeletons were made in three different ways: there were some that were made of calcium carbonate, some of the material called chitin, and others which made the skeleton from tiny sand grains from the sea bed. Foraminiferids can be found from Ordovician to recent times, although it has been suggested that some forms may have existed in rocks of Cambrian age.

How long has man been walking the Earth?

Ever since the philosopher and naturalist, Charles Darwin, and his contemporaries including Wallace, suggested that man had evolved from apes, there has been a continuous, and almost frantic quest for evidence of our ancestors. Of course, where we come from has always been a fascinating question, but it is only comparatively recently that systematic searches have been made for traces of ancient men by such eminent scientists as L. S. B. Leakey and still more recently by his son, Richard. As you might imagine, the idea that man came from monkeys brought with it a great deal of public outcry, and in fact, modern research has shown that man did not descend directly from apes but that the apes and man evolved from a common ancestor.

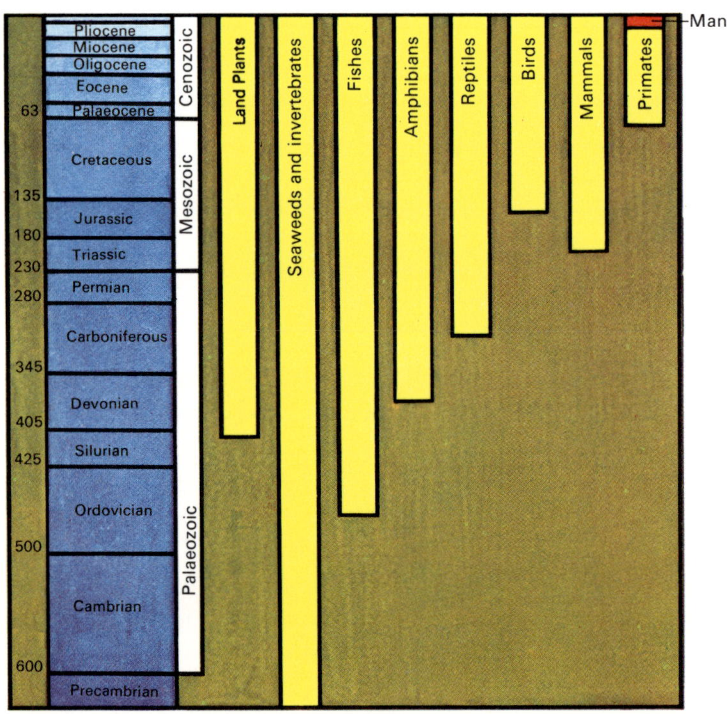

Right:
A representation of the main groups of life forms that have appeared throughout the Earth's long history.

As you know the first thing that monkeys, apes, and man have in common is that they all have backbones, so that they are all vertebrates. Many characteristics in man have arisen from living in trees. We share them with other primates, as monkeys, apes and man are known. If you suddenly found that you had to live in a tree, what do you think the most useful adaptations would be? There would not be much point in having hoofs like a horse which help the horse to run fast. You would need to be able to hold on to the branches, however, as our hands are well able to do. If you wanted to jump from branch to branch, you would have to be able to judge distances. You would need stereoscopic vision, that is, two eyes positioned at the front of your head, not at the sides as in a fish. You are unable to smell the traces of scent that a dog can, but you do have a much enlarged brain. It is the growth of the brain that has enabled man to leave the comparative safety of the trees and compete with the other ground-dwelling animals that are stronger, have better hearing and sense of smell, can run faster and have warm coats to protect them from the extremes of climate.

As you might expect, because man and apes have developed from a common ancestor, it is not easy to tell when the remains of the primates that have been found are of men-like apes or ape-like men. But between ten and fifteen million years ago a creature which has been called *Dryopithecus* lived in parts of Africa, Asia, and Europe. This may have been the ancestor of man and the other apes. From this point, the two groups diverged, until the first creature that could be called man arrived. The most up-to-date research indicates that this may have been as long as four million years ago.

Right:
It is estimated that *Homo habilis*, a primitive man able to use tools, lived a little less than two million years ago.

Homo habilis

Below:
Peking-man, apparently able to use fire, probably lived between 400,000 and 500,000 years ago.

Who were man's ancestors?

Above:
An impression of a very early primate, *Notharctus*.

We have briefly looked at some of the developments which have occurred in man's evolution from the ape-like creatures of prehistory to the learned, technological *Homo sapiens* of today. But perhaps it is worth looking in more detail at some of the animals which have been proved by the investigators to be related in some way to ourselves.

The first creatures which have been identified as man's forebears were certainly not human. About two million years ago, ape-men known as Australopithecines – southern ape-men – inhabited parts of south and east Africa, as their remains indicate. Some of these ape-men lived in the open plains and were small, nimble creatures, well adapted to their life in these conditions. The other group lived in the forested regions, and consequently were much larger, stronger animals.

It has been suggested that these southern ape-men were among the first primates to use tools made of bones and teeth. It may be, however, that the remains of animal bones and teeth which have been found associated with the Australopithecines were just the leftovers from meals. But whether or not they used tools, the teeth of the plains' ape-men certainly suggest that they might have eaten meat and, of

Right:
Here you can see how the skulls of some of the early men developed and this illustration gives an interpretation of their appearance.

course, weapons would have been particularly useful in hunting and killing their victims. It is likely that their relatives living in the forests had a more vegetarian diet.

About one-and-three-quarter million years ago a primate existed, also in east Africa, which has been proved to belong to the genus *Homo*, that is it was the true man. There may yet be finds to come that will indicate an even older date for the origins of a true man, but as yet this is the first creature that has been proved to have made tools and weapons in definite shapes and sizes. The evidence provided by finds of bones from feet of these early men show that they probably walked on two feet like we do, rather than on four feet like an animal such as a dog. *Homo habilis*, as it is called, was probably the first primate to have been able to stand upright.

From this point onwards man seems to have made very rapid progress, and there is fossil evidence of a number of primitive men, but perhaps the most famous of all is the man whose bones have been found in eastern Europe particularly, *Homo sapiens neanderthalensis* – Neanderthal-man. This knock-kneed, hunch-backed creature is very closely related to modern man, *Homo sapiens sapiens*, but was probably not directly related to him and may have evolved from another form of *Homo sapiens* such as *Homo sapiens steinheimensis*.

Above:
The skull and upper limb of a fossil ape, *Proconsul africanus*. This fossil ape may have lived as much as twenty-five million years ago.

What kinds of tools did early man use?

Some animals and birds make use of tools. There are some good examples among the birds. For example, the song thrush commonly feeds on snails which it opens by flying up with the shell in its bill and dropping the snail on a convenient stone. The bird often uses the same stone on a number of occasions so that it may become littered with broken shells. The stone is often referred to as a thrush's anvil. The members of the shrike family often make use of thorns on which they impale their prey. But man more than any other animal has taken advantage of his large brain capacity to increase his power to hunt, kill, build his homes, and so on.

It is man's ability to hold things in his hand which has enabled him to make such use of tools, where other animals have had to rely on their strength and agility, their teeth and their claws. Man's first tools and weapons were stones which he had picked up. He chose stones that had been worn and sharpened by the weather into shapes that could be used as clubs or for cutting and scraping. As you might imagine the right shaped stones were not always easy to find, so that in time man learned to fashion primitive tools to a constant design that suited his purpose.

Many of the early men were carnivorous, that is, they fed largely on meat. This meant that a lot of the stones that have been positively identified as prehistoric implements were designed to kill animals, and then to scrape the flesh off the bones of the victim. Tools of prehistoric man could be made

Right:
Neanderthal-man. He is using a stone implement, like the one illustrated, to scrape a hide.

Above and right:
These are the kind of flint hand-axes that primitive man might have used about 400,000 years ago.

of bone, tooth, or more frequently stone, and in particular, flint. Flint was chosen because it breaks to give a flake with a very sharp, hard edge that could be used as, say, an axe. At first, these primitive weapons would be held directly in the hand. Eventually, however, man learned how to attach these axe-heads to a wooden shaft, and the power of the weapon was increased.

As he became more experienced in making stone tools, the tools became more and more advanced in their design so that soon he was making blades, scrapers, and even arrowheads. Arrowheads were a great advance because they meant that for the first time man could kill his prey without putting himself to the risk of approaching what could be a very dangerous animal. Of course, these weapons could also be used to wound and kill other men.

How should you collect fossils?

You have seen that during the long history of the Earth, there have been many species of animals and plants that have existed some which are still alive today, some long since extinct. Fortunately, the existence of most of these species has not gone unrecorded even though they may have vanished long before the appearance of man. We have fossils to thank for a comparatively clear picture of the life of the past.

Right:
Here are some items of equipment which will be useful when collecting fossils.

Collecting fossils can be fun for all sorts of reasons. You may be interested in seeing at first hand the remains of creatures very different from those common today. You may want to collect fossils as an aid to other geological pursuits such as stratigraphy. You might just like the fascinating shapes that can be found. But whatever your reasons, it is important to go about fossil hunting in the right way, in the best places, using the most useful equipment.

As you know, some rocks are more likely to contain fossils than others. For example, it would be remarkable indeed to find a fossil in serpentine rock of the Lizard in Cornwall, but in Devonian limestones of Torquay not too far away, it would not be at all unusual. Before you begin your searches, consult a geological map of the area and the regional guide published by the Institute of Geological Sciences. A visit to the museum will also furnish valuable information. Then choose the best areas from what you have learned, bearing

in mind that cliffs, quarries, road and rail cuttings, river banks, and even the beach provide the best sites. Remember that these areas are dangerous, and for quarries it is usually necessary to obtain permission first. Make sure you are accompanied by someone who is experienced, and follow the country code.

For the best results you will need some equipment. For the most part, equipment is dealt with in another question, but other items should include newspaper to wrap your finds in, a pen with waterproof ink, and a notebook. You must remember always to mark the find with an identifying number which you have recorded in the notebook together with when and where it was found, including the rock it was found in and its exact location within that rock. It cannot be stressed too strongly that great care must be taken when removing any fossil from its containing rock. Many valuable specimens have been lost by over enthusiasm with a geologic hammer. Remember that a fossil can be very delicate. Think how long it has been resting there and think how quickly it could be destroyed. If you have difficulties, take home the whole rock so that you can remove it at home. If you are searching among clays or soft sands a trowel or a stout pocket knife will be more useful than a hammer, and for tiny fossils a selection of sieves will be valuable.

Left and above: This young geologist is carefully removing a fossil specimen which will be taken home and catalogued.

Fusus
(Gasteropod–Cretaceous to Recent) 1 cm

Micraster
(Echinoid–Cretaceous)

1 cm

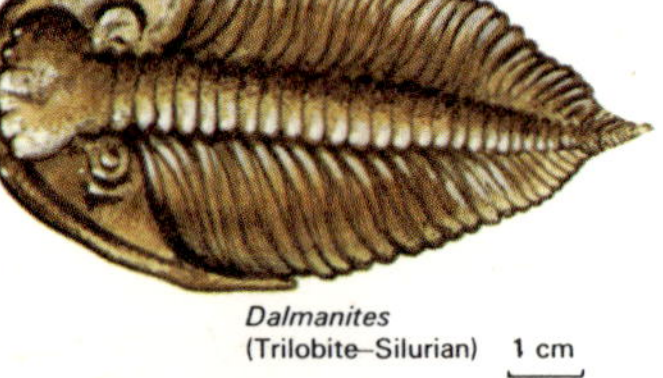

Dalmanites
(Trilobite–Silurian) 1 cm

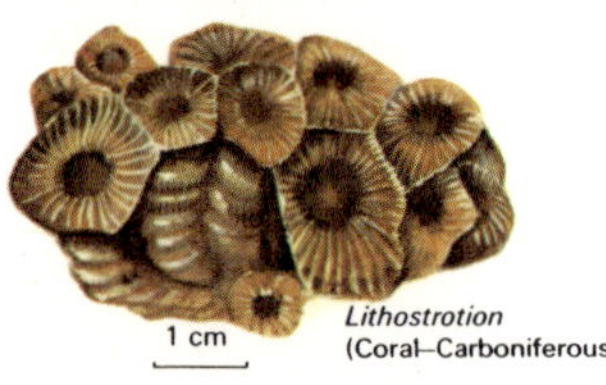

1 cm *Lithostrotion*
(Coral–Carboniferous)

Dactylioceras
(Ammonite–Jurassic) 1 cm

Alethopteris
(Plant–Carboniferous) 1 cm

Above:
Here are some of the fossils that can be found in the geologic time scale.

How could you build up a fossil collection?

Having gone to all the trouble of finding the best collecting areas, choosing the right equipment, and actually going out into the countryside, perhaps in inclement weather, to collect your specimens, you will probably want to make the best of your finds by building up a properly catalogued collection of well-prepared fossils which you will be proud to show your friends. Too many valuable items have been wasted by collection and then being allowed to collect dust, decompose, and finally to be thrown away.

The first thing to do is to prepare the fossil properly. This will depend mainly on two things: what the fossil is and what it is made of; and the type of rock in which it is contained. For example, you may have been lucky enough to find a recognizable piece of bone which is embedded in a limestone. If you have done any chemistry, you will know that dilute hydrochloric acid will slowly dissolve limestone but will leave the bone (made of calcium phosphate) unharmed. This, then, is a useful way of removing a specimen of this type from the rock. But remember, that any acid can be very dangerous and you should never undertake any acid preparation without strict supervision. Lack of care could result in serious burns or even the loss of an eye. You should also consult an expert to make sure that the fossil you wish to prepare will be unaffected by the acid.

Specimens embedded in most clays, chalk, and some sandstones can be removed by carefully scraping the rock away with a pointed instrument such as a large needle. Again take care. It is all too easy to slip and cut your hand or break the fossil you are trying so hard to remove. Sometimes, the gentle use of a hammer and cold chisel will effect the removal of a specimen in a harder limestone. But there is a good chance that the fossil might break. On other occasions, it may not be possible to completely remove the fossil from the rock. If in doubt, do not try, but consult your teacher or someone at a museum. It may be possible just to scrape away some of the rock so that the fossil stands out in relief. Many fossils will also benefit from being coated with a special varnish to prevent them decomposing.

The final thing is to arrange the collection. Place each fossil in a small cardboard tray together with a label stating when and where the fossil was found and if possible, the exact scientific name. There are many guide books available to help you to identify your fossils. For example, the British Museum publish three inexpensive identification guides to

Above:
This illustration shows the best way of arranging your fossil collection.

fossils that can be found in Britain. Do not forget to write a catalogue number on the fossil and on the label, and enter it into a notebook.

Finally you can arrange these trays in drawers either by grouping, say, all the ammonites together, and so on, or by grouping together all the fossils that have come from a particular bed of rock. You may even be able to do both if you have enough specimens.

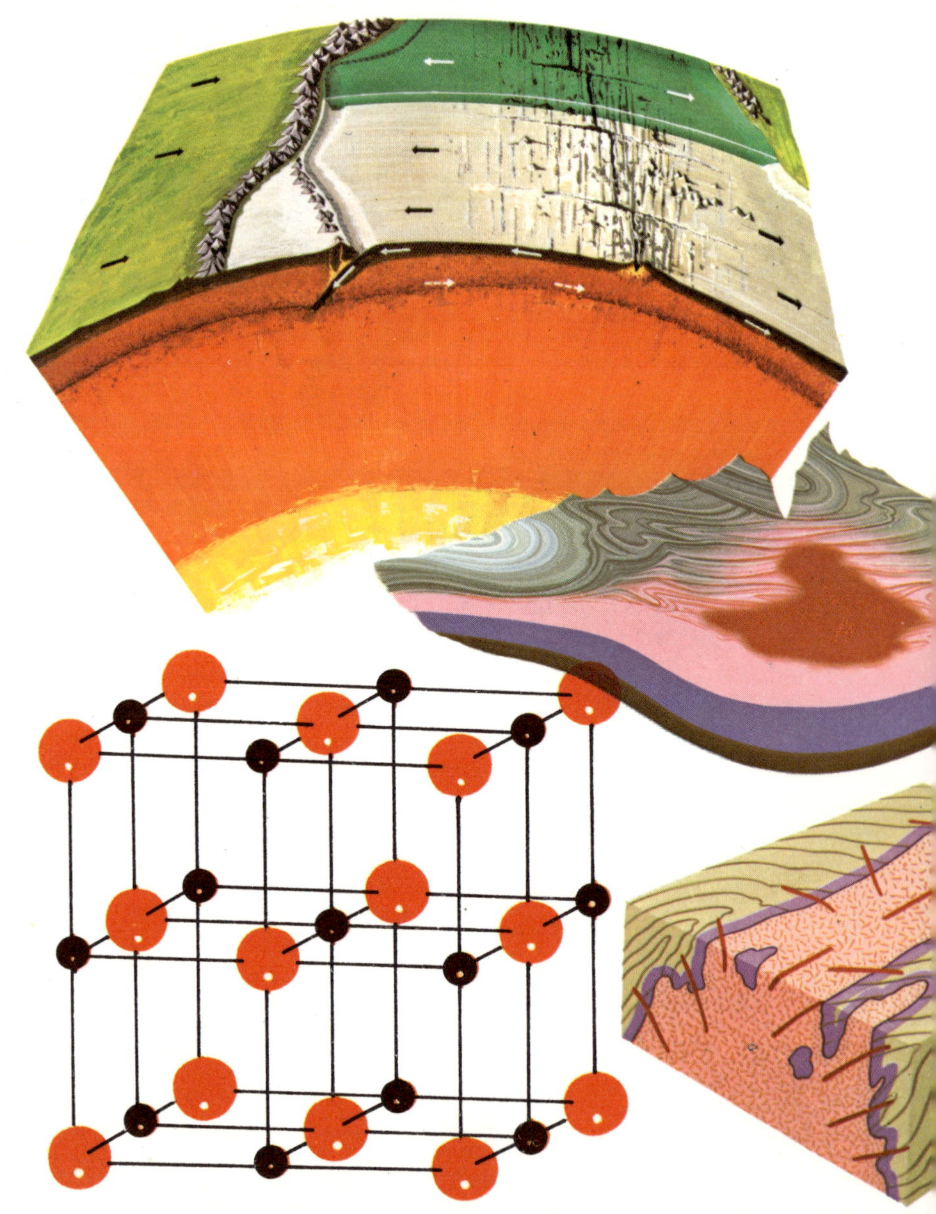

What do you know about Crystals, Minerals and Rocks

What kinds of rocks are there?

If you have been to any of the places where rocks are exposed at the surface of the Earth, such as in cliffs and quarries, or in mountainous areas, you may have already noticed that rocks come in a variety of forms. Even if you have only had the chance to examine the pebbles in your garden, you will probably have seen that the colour, shape, and hardness of the pebbles can vary quite considerably. Remember that soft, white, crumbly chalk and hard, black basalt are both rocks. Even soft, sticky clays are considered to be rocks.

Right and far right: Some typical igneous, sedimentary, and metamorphic rocks.

It is important for geologists to be able to pass on information about any particular rock in as brief and accurate a way as possible, so that geological information can be accumulated. In scientific language a rock could be described as *an acid, igneous rock containing more than 10 per cent quartz, and in which the alkali feldspars predominate over the calcic feldspars*. But this is very long-winded and not easy to understand – it would be enough to say that the rock is a *granite*.

It is possible to group all the hundreds of different rock types in the world into families to give an idea of the origin and/or the make-up of those rocks. Every rock can be given a name to convey more particular information. In fact rocks are usually divided into three major groups on the basis of the way in which they were formed. Rocks may be *igneous*, *sedimentary*, or *metamorphic* in origin. The word igneous comes from the Latin *ignis* meaning fire and immediately gives an idea of how these rocks were formed; that is, that the great heat inside the Earth has been involved in their formation. Igneous rocks are formed by the cooling of *magma*, or melted rock which has been made more fluid by gases. A granite is a typical igneous rock.

As the name implies, *sedimentary* rocks are formed from compacted, hardened, and cemented sediments such as sands to form sandstones, clays to form shales and limey oozes to form chalk. As you have seen the sediments arise from the wearing away of other rocks. *Metamorphic* rocks are rocks that have been changed by heat and pressure. A slate is the metamorphic version of a shale.

What are rocks made of?

You could answer this question by just saying that rocks are made of minerals. But then there is the question: what are minerals made of? And the answer might be that the basic component of a mineral is a crystal, and so on. In the end we see that rocks are made up of the elements. You have read of the elements in a previous question, and they include oxygen, silicon and aluminium.

Rocks are made up of materials formed by the elements joining together in chemical combination. Some elements, such as gold, do not need to combine with other elements and can exist alone in the Earth's crust. But silicon, for example, can combine with oxygen in the proportion of two atoms of oxygen to one of silicon to form the compound called silicon oxide or silica. In nature silica may be found as the mineral called quartz. Quartz may occur in a variety of colours depending upon the impurities it contains, ranging from black or pink to glassy clear. You might find quartz as irregularly shaped broken lumps, but if you look at the pieces closely, they are more likely to have straight edges or may end in a pyramid shape. In fact, if the conditions in which the quartz formed were perfect, a crystal might have formed. What you see is the result of many crystals growing and interfering with each other.

Right:
The metal aluminium, which occurs in combination with other elements in rocks, is a good conductor of heat and electricity. It is used for electricity cables and cooking pots. Aluminium is becoming more difficult to find in workable amounts in the Earth's crust.

A crystal is a symmetrical solid. It has flat surfaces that have a direct relationship to the arrangement of the atoms in the crystal. A mineral may then be described as a solid, inorganic substance occurring naturally in the Earth's crust, and having a definite composition and a structure which is consistent throughout. In this definition the word inorganic may be puzzling. It is a chemical term implying that the rocks do not contain the element carbon in the types of combinations which are normally associated with living creatures, or their basic materials.

This brings us back to the first definition in this question: that a rock is a mass of minerals. Unfortunately, nature is rather fickle and as you will find in many of the other natural subjects, the scientific definitions do not always tell the whole story. Some rocks have been formed by the action of living creatures and still others, such as volcanic glass, are not made up of crystals at all.

Right:
An igneous rock, such as granite, is made up of the minerals quartz, mica, and feldspar.

What shapes of crystals are there?

We have already explained the way in which rocks are made up of minerals, and that minerals take the form of regular solids called crystals that may come in a great many different shapes, hardnesses, and colours. Naturally, you would expect that if you wanted to know more about rocks, you would need to understand more about the way in which crystals develop. The scientific study of crystals is known as crystallography.

The shapes of crystals are very important to anyone wishing to identify a mineral, because these regular forms exactly express the way in which the atoms of that mineral are arranged. This means that if you are able to recognize the shape of the crystal of a mineral, then, with the help of a number of other properties, you could work out what the mineral was by a process of elimination. Other important properties are weight, hardness (which in this case means the resistance to scratching), and so on.

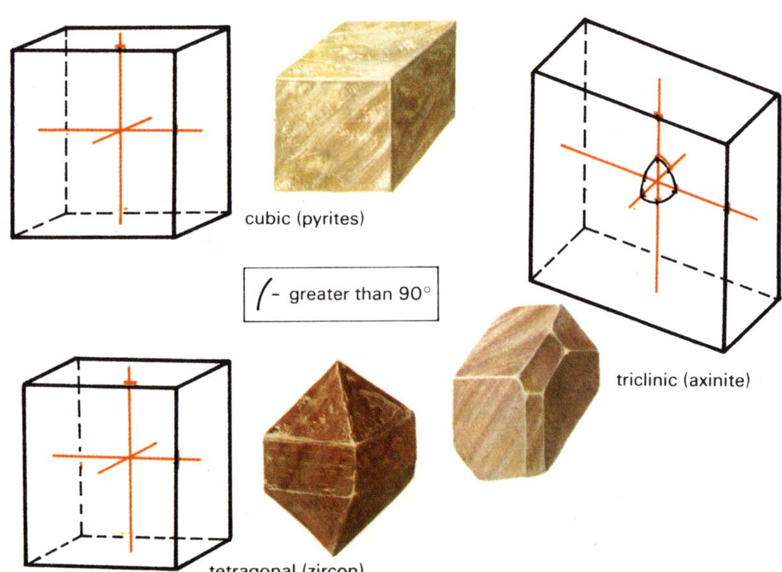

Right and far right: The seven crystal systems and some common minerals that crystallize in each of them.

The most important property, then, in recognizing crystalline materials is *symmetry*, that is, the regularity of the shape. You can try an experiment in symmetry for yourself. Take a fairly large potato and peel it: Be very careful not to cut yourself with the knife. Then cut the potato into the shape of a cube; that is, a dice shape with six square faces and sharp corners. Some minerals such as iron pyrite (fool's gold) often develop as cubic crystals. Now cut off the corners of the potato cube and you will have a shape with eight more faces. If you continue to make these eight faces larger and larger, you will eventually cut away the six original faces leaving a shape with eight faces only called an octahedron.

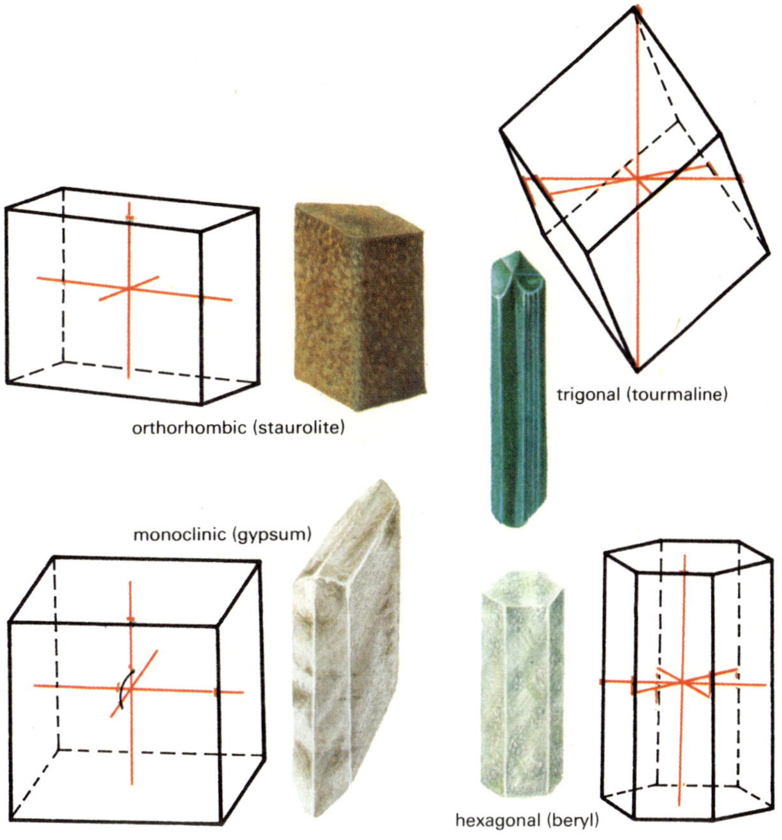

You have now shown for yourself that three seemingly different shapes are related to each other. These relationships occur throughout the world of crystals, so that of all the many crystal shapes there are only seven crystal systems into which they can all be classified. You have already seen one of these systems, the cubic system. The other six are given the names tetragonal, orthorhombic, hexagonal, trigonal, monoclinic, and triclinic. If you can place any crystal that you find into one of these systems then you have gone a long way towards identifying the mineral. With some crystals it is quite easy, but with others it does require considerable practice. We shall explain how to recognize some of the simpler crystals in the next question.

How do you recognize crystals?

In the previous question, we mentioned that there are seven crystal systems. Knowing their names is of little value unless you know how to place a given crystal in its system, and then some of the minerals in each system. The way in which you place a crystal in its class is really quite simple and depends upon what are known as *elements of symmetry*. In fact, each crystal system is actually defined by certain elements of symmetry.

Look at your potato again – the one that you have cut into an octahedron. Hold it between your finger and thumb so that your finger is at one of its points and your thumb at the one exactly opposite. Now turn the potato slowly about these two points, and you will find that each edge or face is repeated by an exactly similar edge or face four times as you turn it through one complete turn. This means that this shape has a fourfold axis of symmetry. You will find that you can hold the potato-octahedron by two other pairs of points and repeat the process so that there are three of these axes. Now hold the potato by two opposite faces instead. When you turn it this time you will find that there are three repeats in one full turn and that there are three other pairs of faces that you can hold. You can see, then that the octahedron has four of these threefold axes of symmetry.

You already know that this shape was cut from a cube. You have now seen how to recognize any cubic crystal. To belong to the cubic system, a crystal must have four threefold axes of symmetry, and you can look for them in exactly the same way as you did with the potato-octahedron.

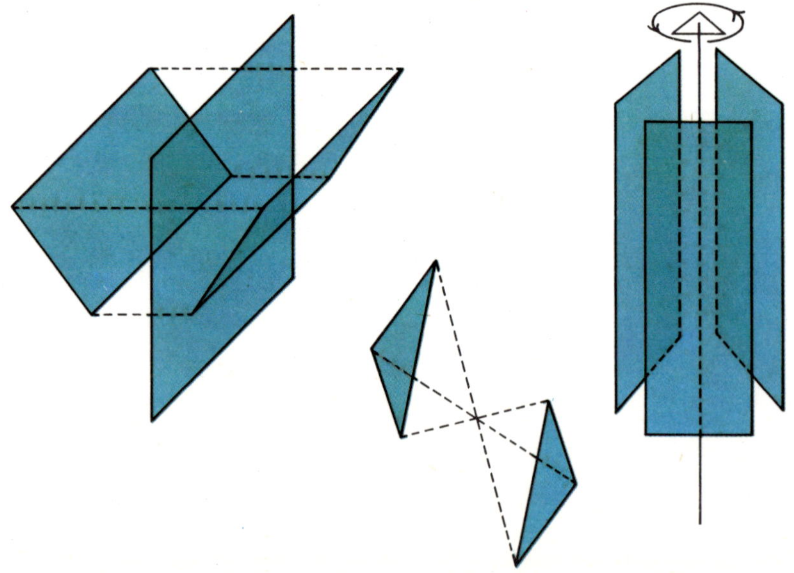

From left to right:
A plane of symmetry, a centre of symmetry, and a threefold axis of symmetry.

Right:
Any crystal reflects the arrangement of its atoms; this simple arrangement of two different atoms is typical of a substance such as common salt.

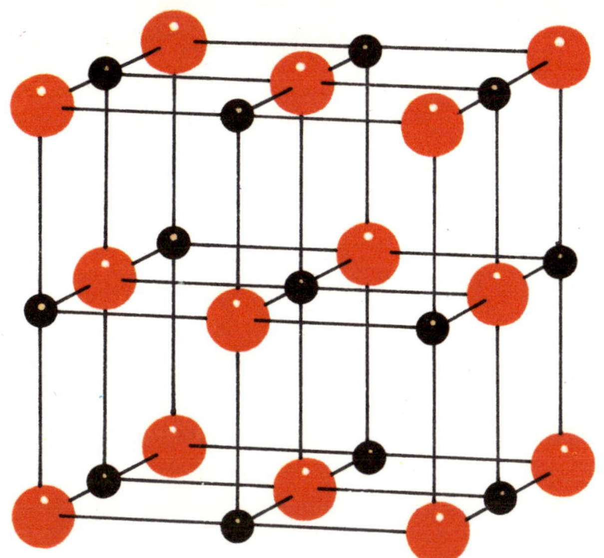

You can soon recognize the other six systems of crystals by working out their axes of symmetry. Each system is defined by a different set of rules. Crystals belonging to the tetragonal system must have one fourfold axis of symmetry and one only. It might also have four twofold axes if it were a very regular example of the system, but there are some that do not. An example of this system is the mineral zircon which is sometimes used as a stone in jewellery. Orthorhombic crystals must have three twofold axes, hexagonal crystals one sixfold axis, trigonal one threefold axis, monoclinic one twofold axis, and finally triclinic with no axes at all.

Right:
Silicon atoms and oxygen atoms may join together in chains of tetrahedra like the one shown. You can see how the chains are packed together into the crystal shape.

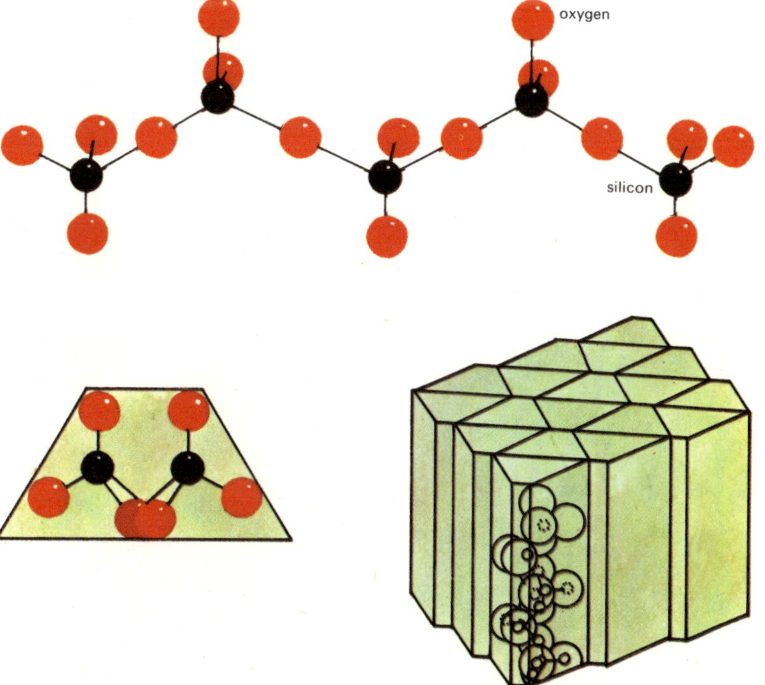

What are the main mineral groups?

We have already explained the way in which the form of a crystal is directly related to its internal atomic structure. You have also seen how the many possible crystal shapes can be grouped into just seven crystal systems. To simplify their study, the Earth's minerals can also be grouped together. The way in which they are grouped depends upon the purpose of the grouping. For example, you might want to group minerals on the basis of the elements which compose them; this would be a chemical classification. On the other hand, you could group together all the minerals of the same colour, or all the minerals which have crystals of the same system, but these groupings would not tell you anything about the ways in which the different minerals were formed.

A simple grouping which is often used by mineralogists (those who study minerals) is into *rock-forming minerals* and *non rock-forming minerals*. Naturally every mineral occurs in a rock, but the rock-forming minerals actually make up the bulk of the igneous rocks. You will remember that igneous rocks are formed by the cooling of magma and they provide the materials for sedimentary and metamorphic rocks.

Right:
Some of the many different silicate minerals that you might come across in rocks.

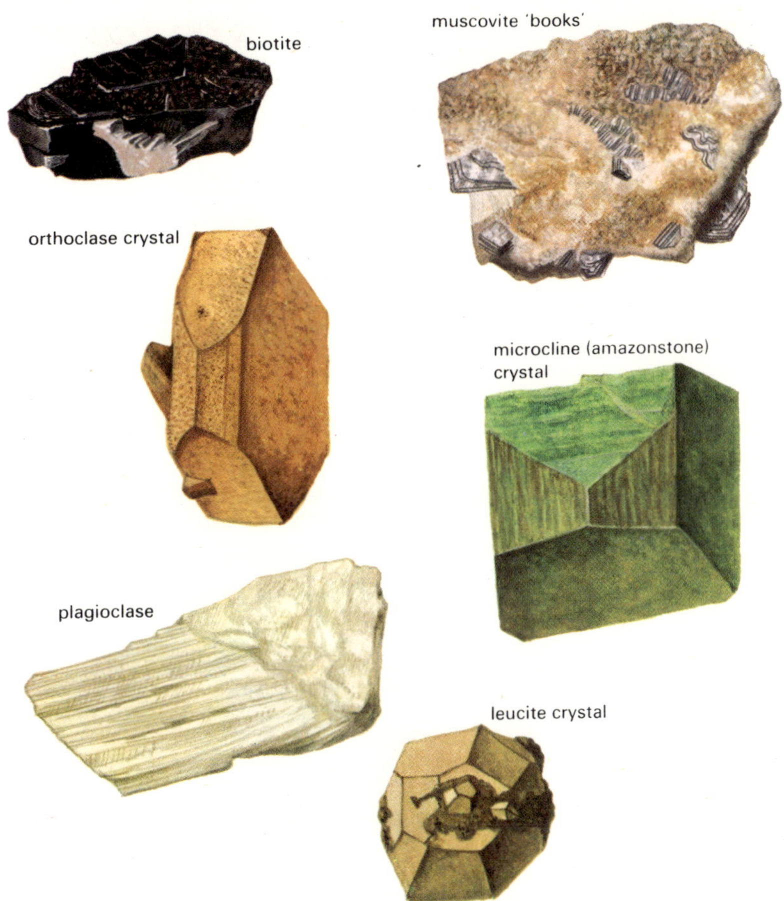

Right:
Here are some more silicate minerals.

Another important grouping is into *silicates* and *non-silicates*, but this is rather more complicated and the details are beyond the scope of this book. A very great simplification of this grouping would be to say that silicates contain silica in combination with other elements and non-silicates do not contain silica.

The silicates are by far the most important of the rock-forming minerals, and are worth more detailed consideration. These silicate minerals may be further divided into groups which have similar internal, or atomic, make-up. To be more scientific, it has been shown by a researcher called Bowen, how members of a group are formed from molten rock at different temperatures. As you know, heat is just one form of energy. Minerals that are formed at very high temperatures, such as olivine, are formed by a great deal of energy which is reflected by the way in which its atoms are arranged. The other main silicate minerals are given the names pyroxenes, amphiboles, micas, feldspars and quartz. The order in which they have been mentioned exactly corresponds to the temperature at which you would expect them to form from the melted rock material.

How do you identify minerals?

When you recognize a person, the recognition is based on your knowledge of certain characteristics of that person. Without your being consciously aware of it, your brain is very quickly looking at the person's height, build, colour of hair, facial characteristics such as colour of eyes, size and shape of mouth, and so on. Similarly, when you wish to identify any mineral, there are a number of properties that you need to look for. Some of these properties can only be examined with the use of a special microscope, and with many minerals they can only be positively identified in this way. But there are a lot of minerals that can be identified quite accurately by carefully looking at them with the naked eye or with the aid of a simple magnifying lens (one with a magnification of × 10 is ideal).

Right:
The mineral fluorite can occur in a variety of colours. The iron minerals limonite and haematite have different streaks.

When identifying a mineral, you do need to know the features to look for. Each mineral has a different internal make-up which is reflected in its properties. You need to work out the crystal class to which the mineral belongs, its hardness (resistance to scratching), specific gravity, colour, lustre, degree of transparency, streak (colour of powder when crushed), and sometimes the taste or the smell. In the case of taste, you should never try to taste the mineral yourself because many are poisonous. You need to be very experienced before using this test.

We have already explained how to establish the crystal form, and this is very important. There is a scale invented by Mohs to test hardness, which indicates by the numbers one to ten the hardnesses of certain reference minerals. A mineral of hardness 3 will scratch a mineral of hardness 2 but not one of hardness 4 and so on. The softest mineral is talc and the hardest diamond.

Right:
These minerals are exhibiting different lustres.

Specific gravity can be estimated if you are experienced, by holding the specimen in your hand and feeling if it is light or heavy for its size. Colour may be affected by impurities and by weathering. Lustre refers to whether the mineral shines like a metal or looks glassy or pearly and so on. The transparency refers to how well you can see through the mineral – is it completely clear or not? To test the streak, simply scratch the sample on a piece of unglazed porcelain and look at the colour of the trace. Another important feature is the way in which the sample breaks – along certain planes or irregularly. These are the properties you look for when you are trying to work out a mineral 'fingerprint'.

Right:
Minerals can break or *cleave* in a variety of ways – these are just a few.

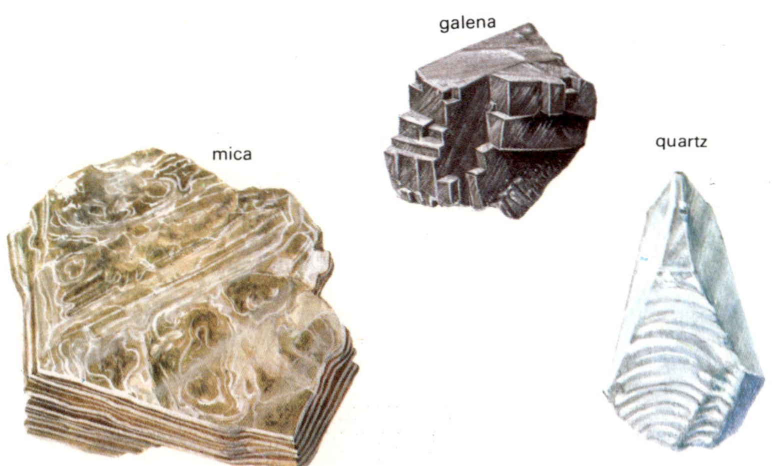

Which rock burns?

All rocks will eventually burn or melt at high enough temperatures. The rock we are referring to burns at quite low temperatures and provides us with many chemicals. It can be very soft and crumbly or quite hard; it can be shiny or quite dull. It can be burnt in its natural state or it can be treated so that it burns without smoke; it is usually black. Of course, the rock is coal! Like oil, coal is often regarded as a mineral, although as you will see, it also has living origins.

Certain parts of the world, such as the bogs of Scotland and Ireland, the American swamps and places like Burma are water-logged. Plants that live there do not decay in the normal way when they die. The bacteria and fungi which break down the plant remains under more normal conditions, cannot live because there is not enough oxygen and certain acids are formed which kill them. Under these circumstances, a jelly-like humus is formed and this soaks into the fragments of wood and bark and so on, to form a material called *peat*.

Peat is rich in carbon, and the chemical reactions which occur also produce marsh gas or methane which burns very readily. Methane is the natural gas which is piped into most of your homes. In more remote places, such as the north-west of Scotland, where peat is still being formed, it is dug up during the summer months and stacked to dry so that it can be used for fires during the long, cold winter.

As the peat builds up year after year, the water is squeezed out, and then with further burying under clays and sands it undergoes a change. It becomes more and more compacted and the gases are pressed out. It becomes harder and richer in carbon until eventually it forms a *brown coal*. It has often

Right:
Coal is deposited as one of a sequence of sedimentary events and the rocks that occur above and below a coal seam, together with the contained fossils, indicate the conditions of deposition in the cyclic sequence.

SEDIMENTS	FOSSILS	ENVIRONMENTS
coal	leaves, stems, and spores of trees	swampy forests
seat earth	roots	
mudstone		surface built up to water level
sandstone locally cross-bedded with irregularities		deltaic
		influx of river sands
siltstone	plant debris	deltaic lagoons
mudstone and shale		delta front or estuarine brackish water
with ironstone nodules	mussels, fish	
mudstone	marine molluscs	inundation by the sea
coal		swampy forests

Right:
Coal can be mined by tunnelling underground or by digging huge, open pits. These pits are called open cast mines.

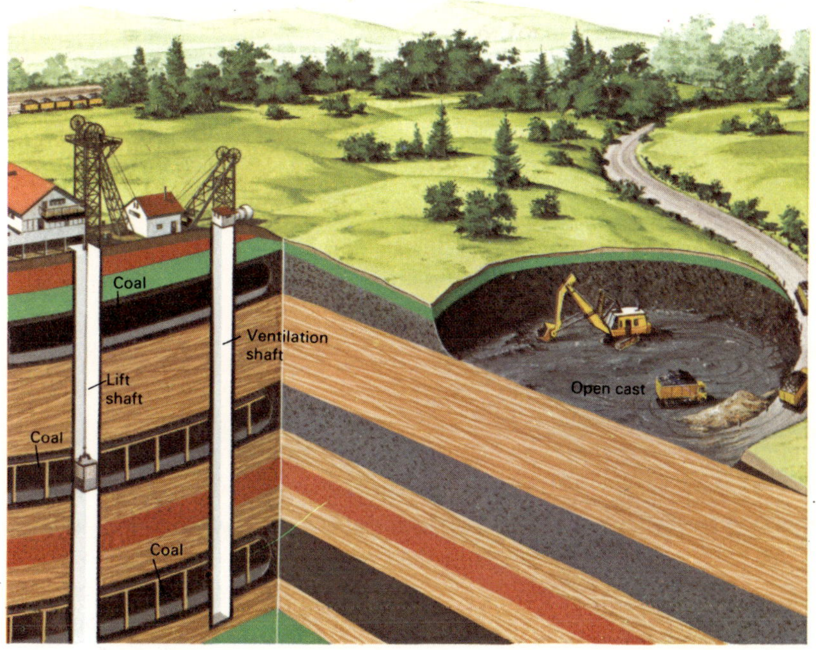

been shown in coal fields that the greater the thickness of rock on top of the coal, the higher the carbon content of the coal. Eventually anthracite results which contains about 95 per cent carbon and is a very high quality and expensive coal.

Most of the coal fields throughout the world are to be found in rocks of Carboniferous age. About 300 million years ago, dense swampy forests occurred because of the very rich plant life which had by then developed on Earth, and it is the remains of these forests that gives us most of our coal.

Right:
This was how an eighteenth century naturalist represented the coal seams of Somerset, England.

Which mineral comes from animals and plants?

We have already given a definition of a mineral in a previous question, and two of the properties were that a mineral must be solid and inorganic. There is one material which is always included as a mineral, however, which does not follow the rule. We have mentioned it before. Can you guess what it is?

The mineral to which we are referring is correctly called petroleum from the Greek *petra* meaning a rock and the Latin *oleum* meaning oil. Petroleum is the name given to all the materials which occur in nature and are known chemically as hydrocarbons. This means that they are composed of combinations of carbon and hydrogen. These materials may be gases, such as the natural gas which now supplies some of the domestic gas in Britain. There are also liquid oils from which we obtain our vital petrol (or gasolene as it is known in the United States). Many other important products are derived from petroleum, from perfumes to plastics. You can appreciate, then, the importance of this mineral to our way of life in the industrialized nations.

From the way in which the question is worded, and from what you have read so far, you will have guessed that the mineral that comes from animals and plants is, in fact, petroleum. Actually, many different origins have been suggested for this mineral. They have included the ideas that petroleum has been produced by chemical reaction, from volcanoes, and from the decay of land plants. But the evidence that has been gathered at an enormous rate

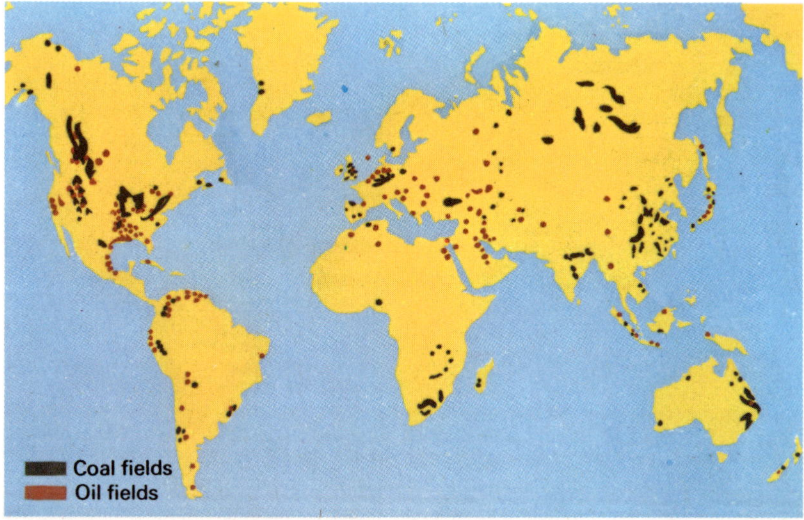

Right:
This map shows the distribution of the world's coal and oil fields.

Right:
Oil, filling the cavities in a porous rock, may become trapped in economic amounts in a variety of ways. Oil is associated with natural gas and water and is trapped by impervious layers of rock.

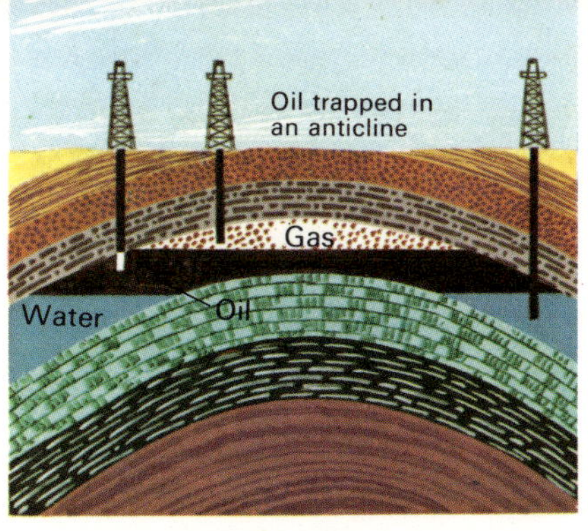

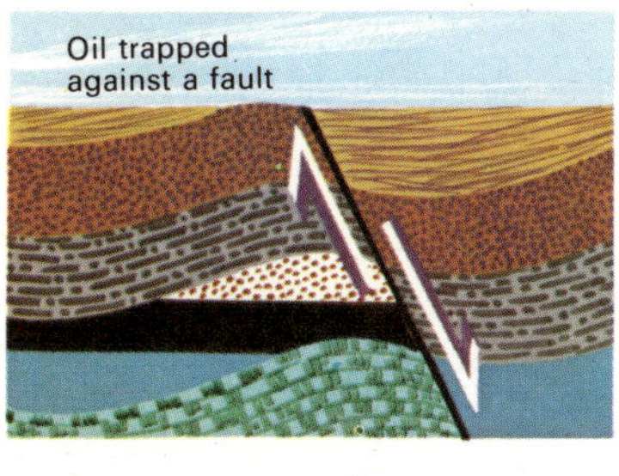

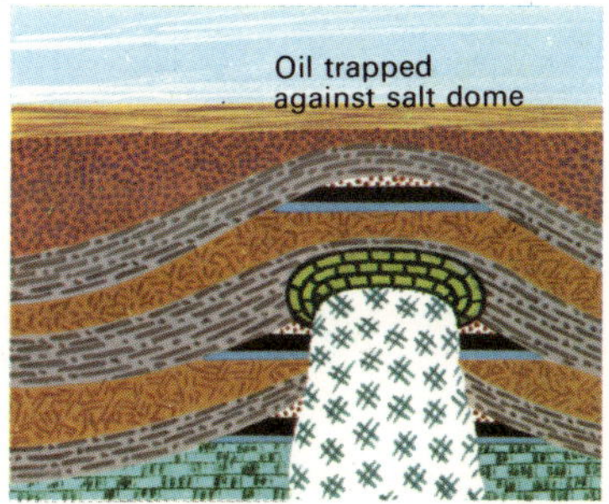

because of the economic importance of oil has shown that none of these ideas is likely to be correct. It is much more probable that oil was formed (and is still being formed today) from the decay of tiny marine plants and animals such as algae and the one-celled diatoms.

The two most important conditions for the formation of petroleum are the activities of certain bacteria and time. It is very important to remember that petroleum products take millions of years to form and concentrate into quantities which can be worked. But in 100 years we have consumed a large proportion of the world's oil reserves. Oil, then, formed as tiny droplets among the particles of muddy sediments at the bottom of the sea where the water was stagnant and there was little air.

Right:
An oil pipeline to carry oil from its source to a refinery.

How can you see through a rock?

This seems to be a strange question to ask doesn't it, but for the petrologist (a scientist who studies rocks) it is vitally important. We have already explained what properties to look for in order to identify a rock, and that a rock is a mass of minerals. If you want to identify a rock you must first be able to work out what the minerals are that compose it.

Once again, the first thing you can do if you are looking at rocks out in the field is to examine closely with the aid of a hand lens and try to recognize the main minerals. You have seen how rocks can be igneous, sedimentary, or metamorphic and, with experience, you can identify the type of rock straight away. It takes more practice to positively state the name of a rock within these groups and in many cases you need to use a microscope.

It is at this point that you need to be able to see through a rock. To do this, it is necessary to make a thin section of the

Below:
Examination of a rock in the field with a hand lens can provide an approximate identification. With many rocks, it is necessary to take a specimen to a laboratory for cutting and mounting on slides.

examine with hand-lens

cut slice with diamond saw

mount on glass slip

grind to final thickness

fix cover glass and labels

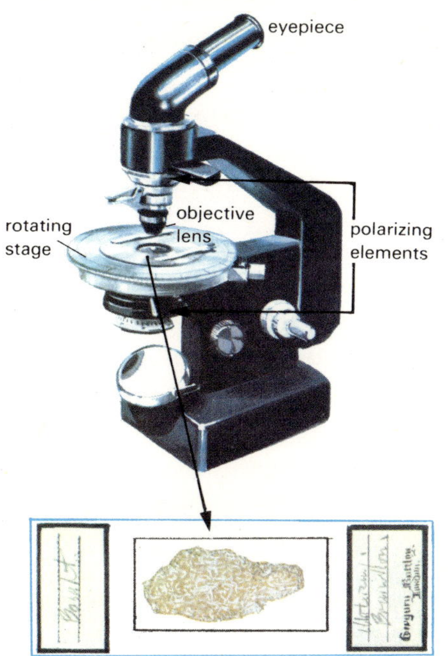

mounted rock thin section

Right:
When the rock samples have been prepared as slides, they are examined with a polarizing microscope for positive results.

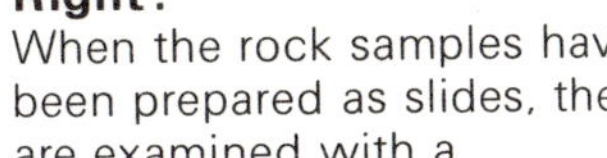

section viewed in ordinary light

section viewed with polarizing elements in place

rock in a laboratory. A piece of the rock is taken and sliced using a special kind of saw, which, surprisingly, will cut through the hardest rock but will not cut the finger of the person using it. This small slice of rock is then glued to a normal microscope slide with a material called Canada balsam. It is ground down until it is exactly 0.03 mm thick. Almost all minerals become more or less transparent at this thickness, and you can see through the rock. But why should anyone want to go to all this trouble?

If the section is placed on a petrologist's microscope it can then be very carefully studied using a special kind of light called *polarized light*. Any light wave vibrates in all directions at right-angles to the direction it is travelling. To polarize it means that all the vibrations except those in one direction only are cut out. This is done by shining the light through a kind of very fine grid.

When polarized light is passed through the thin section and then examined in the microscope, each mineral shows certain characteristics which are peculiar to that one mineral. More properties can be examined if the polarized light which has passed through the section is passed through another grid with its bars at right angles to the first grid. Using this technique, the size and shape of the minerals can be looked at. Petrologists can also measure the angles between the planes along which the mineral breaks. In other words you can get a great deal of information about a rock by being able to see through it.

When is a mineral a gem?

insect in amber

Above:
Amber is actually fossil tree resin so that it is not strictly a gem. It is often used in jewellery, however.

A mineral can claim the title of gem if, in its natural state or when cut in certain ways, its properties are thought to be beautiful. Gems are those minerals which are prized throughout the world for their eye-pleasing appearance.

Some gems are very precious indeed. Diamond, for example, has exactly the same chemical composition as graphite, the soft lead-grey material which is used in pencils. Why, then do we value diamond so highly? There are two main reasons. Firstly, diamond has a very high *refractive index*. This means that although diamond is glass clear, it tends to bend the light passing through it much more than glass does. If the diamond is cut in certain ways, it gives a delightful play of colours which no other gem can rival. Indeed, this is the sole reason for cutting a diamond. It is scarcely recognizable in its raw state, and it would certainly be hard to believe that an uncut stone and such gems as the famous Kohinor diamond were the same material. Diamonds are also the hardest known mineral. This has led to its wide use on the cutting edges of oil drill bits, and on saws for cutting the hardest stones or metals. Now, however, diamonds suitable for industrial use can be manufactured artificially, but nature is still needed to furnish us with gem quality stones.

Another property which gives a stone its value is rarity. For example, even if the correct materials are present, it takes conditions of extreme heat, and pressures equivalent to being buried as much as 120 kilometres beneath the surface of the Earth to form a gem.

Other true gems are the deep, green emeralds, the transparent red ruby, and the delicate to deep blue sapphires. Emeralds are a variety of a mineral called beryl which occurs associated with granites. The ruby is a type of corundum

Right:
Some of the most beautiful and valuable gems.

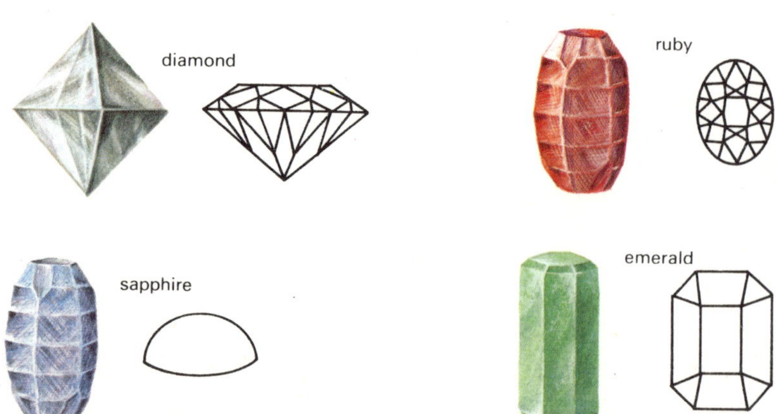

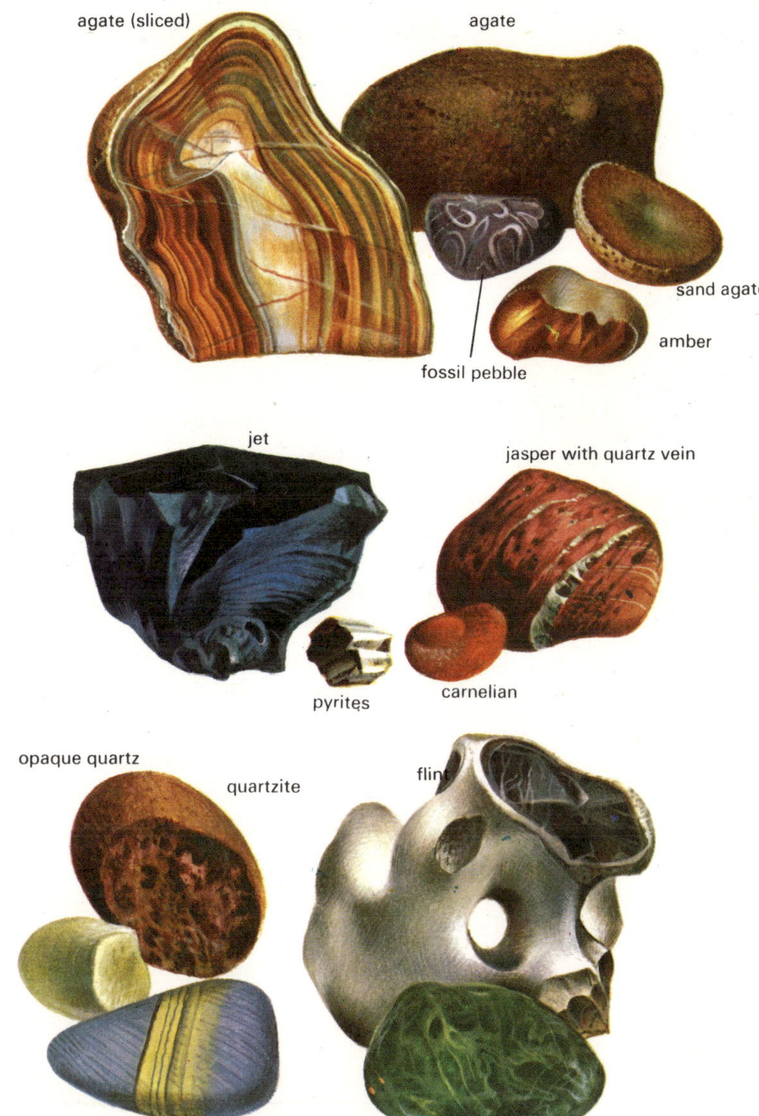

Right:
Semi-precious stones like these can make very attractive jewellery, although they do not have the qualities of hardness, colour, and so on, of the true gemstones.

which you may know as the hard stone that is used to sharpen knives and other cutting edges. Sapphire is also a variety of corundum. All these gems are rare, resistant to scratching, beautiful in colour, have an attractive lustre and can be cut to further enhance their beauty.

As well as these very valuable stones there are other less valuable, but in some ways just as attractive, stones that are known as semi-precious. These tend to be less hard or brilliant in colour and lustre, and rather more common. Many varieties of quartz are semi-precious, such as citrine, amethyst and the non-crystalline opal; the banded ones such as onyx and agate are also very pleasing. Even volcanic glass called obsidian, or the type of soft coal usually called jet, can be polished and are often used in jewellery.

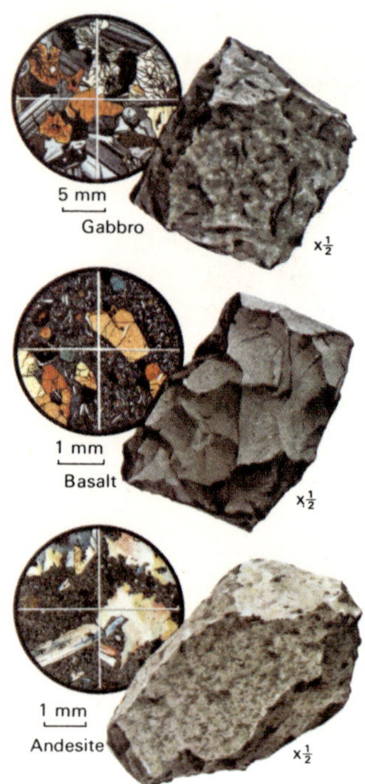

Above:
Some typical igneous rocks viewed in hand specimen and under the microscope.

How are igneous rocks formed?

You have already seen that igneous rocks are the result of the cooling and crystallizing of melted rock material or *magma*, but this is not the whole story, nor has it always been thought to be true. During the first days of the study of our planet, geologists would only believe that those rocks which they could actually observe as lava spewing from volcanoes were actually formed in this way. Many rocks, such as the granites and basalts which petrologists now know to have a fiery beginning were once thought to have formed in the sea. These *Neptunists*, as they were called, could not believe that the crystals which they could clearly see made up many of these rocks, were able to form other than by crystallizing out of a solution containing the correct chemicals, in other words the sea. In fact, these Neptunists went to even greater lengths to defend their ideas. They stated, for example, that the eruption of a volcano was due to the burning of coal beneath the volcano. Unfortunately, they ignored the fact that many of the world's active volcanoes, and many of the places where basalt rock can be found, are nowhere near any coal deposits. Of course, these ideas were to some extent confirmed in the minds of their proposers, by association with the flood described in the Bible. The fact that the whole world was supposed to have been inundated by the flood would account for there being basaltic rocks in areas far removed from the sea.

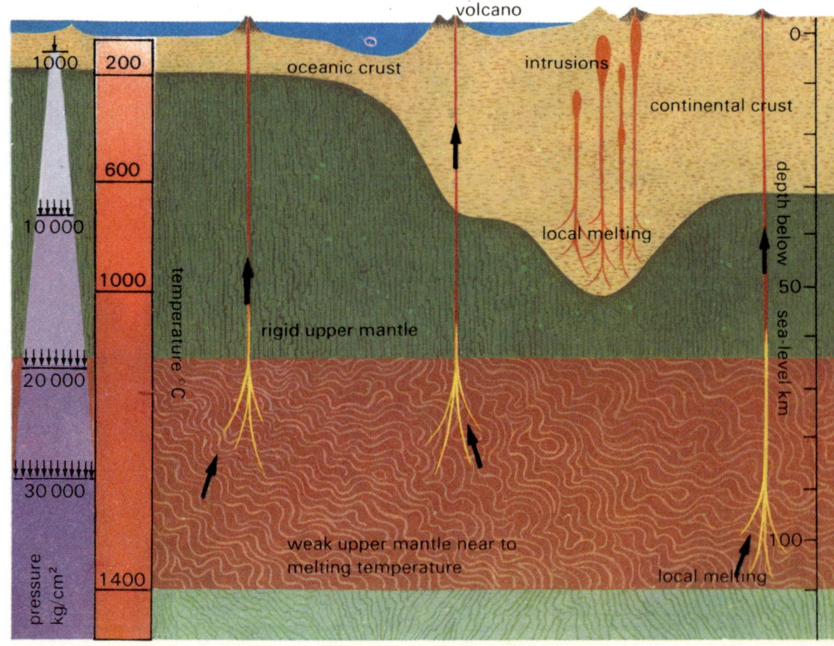

Right:
The magma, which eventually cools to give igneous rocks, forms in different zones in the Earth and at various temperatures and pressures.

Right:
This is how Nicolas Desmarest mapped a lava flow in the Auvergne, France, during the eighteenth century.

Below:
More examples of typical igneous rocks.

Eventually, however, Nicolas Desmarest proved beyond doubt that the ideas of the Neptunists were wrong. In the Auvergne district of France he found some basalt rocks beneath which were cindery lava and soil that had clearly been baked by intense heat. He followed these bands of rocks and found that eventually they could be traced into the crater of a volcano.

From a first glance, then, at some of the world's rocks which are now known to be igneous, you might quickly recognize four distinct characters. You might notice that some of these rocks are composed of many many tiny crystals, that is, they are fine grained. Some are made up of much larger crystals, they are coarse grained. You may also observe that some of the rocks are definitely light in colour and others very dark. Of course, there are many combinations and variations of these main types.

Generally, however, fine grained rocks are the result of the cooling of lavas emanating from volcanoes, and the coarse grained varieties arise from large bodies of molten rock material which cool much more slowly at great depths in the Earth's crust. It is this difference in the rate of cooling which leads to the difference in grain size. The colour is an indication of the chemical make-up of the rock, but more about the composition of rock later.

Why does a volcano erupt?

By now you should be aware that in the bowels of the Earth all is not still. On the contrary, it is very active. One obvious example of the Earth's internal workings is the activity of volcanoes. There are many famous instances of a volcano erupting and spewing white hot, molten rock from its mouth; if you have been to Pompeii or just seen pictures of it you will realize how a whole city and most of its inhabitants can be engulfed in the fiery ash suddenly erupting from what was thought to have been a dead volcano. A more recent example was the eruption on Tristan da Cunha in 1961 when the whole population of the island had to be evacuated, some never to return. Perhaps the most famous of all volcanoes is Krakatoa. This volcano had remained dormant for 200 years, when on 27 August 1883, the volcano exploded with such force that the whole island was ripped apart. The noise was heard almost 5000 kilometres away in Australia, and a tidal wave 36 metres high was sent across the sea to Java, killing 36,000 people. You can understand, then, that it is a very great force that makes volcanoes erupt.

Volcanoes were once thought to be burning mountains and it is not difficult to understand why. But nowadays, a volcano is usually defined as a direct feed from a magma chamber in the interior of the Earth to the surface. Remember that a magma is any hot material within the Earth which can flow and penetrate into or through the surface

Right:
When Mount Vesuvius erupted in AD 79 to bury the city of Pompeii, many of the 20,000 inhabitants were killed and their remains can be seen preserved in the positions in which they died, trapped by the hot ash and pumice.

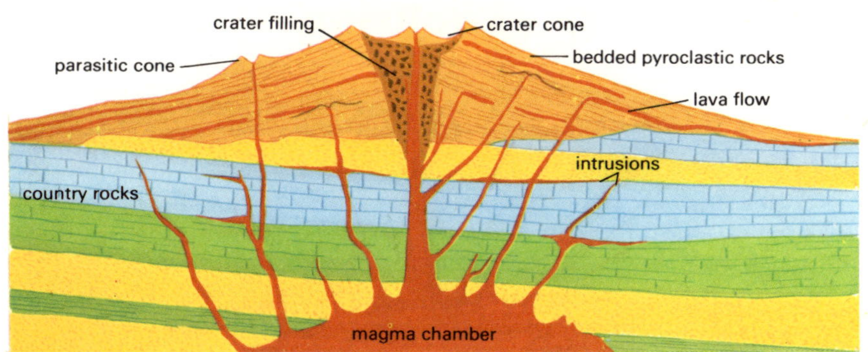

Above:
A section through a volcano known as a *stratovolcano*.

Above right:
Molten lava flowing down a slope.

rocks. A magma, then, is usually thought to consist of a kind of 'porridge' of solid rock material, liquids and semi-plastic substances which are made to flow by the lubrication of hot gases under great pressures.

Two main causes are responsible for the birth of a volcano. Explosive eruptions may be caused by pressure resulting from the build up of gases within the Earth or a crack in the crust can be opened up forming a passage or *vent*. Volcanic lava can then flow through it.

It is important to remember that the classic cone shaped body which you might immediately think of as a volcano, only forms as a result of volcanic activity. It is the pile of ash and cooled lava that has flowed or been thrown out of the original fissure or crack. A volcano may erupt only once, and it may not emit enough material to build up the well-known cone. This cone usually occurs when there are successive eruptions from the same vent.

What kinds of volcanoes are there?

You have seen the ways in which volcanoes erupt, and we have already said that volcanoes are not always the simple cones that one usually associates with outpourings of red hot, molten rock. In fact, eruptions vary a great deal in the type of products that are spewed out, in their intensity, and in the form which the volcanic vent takes.

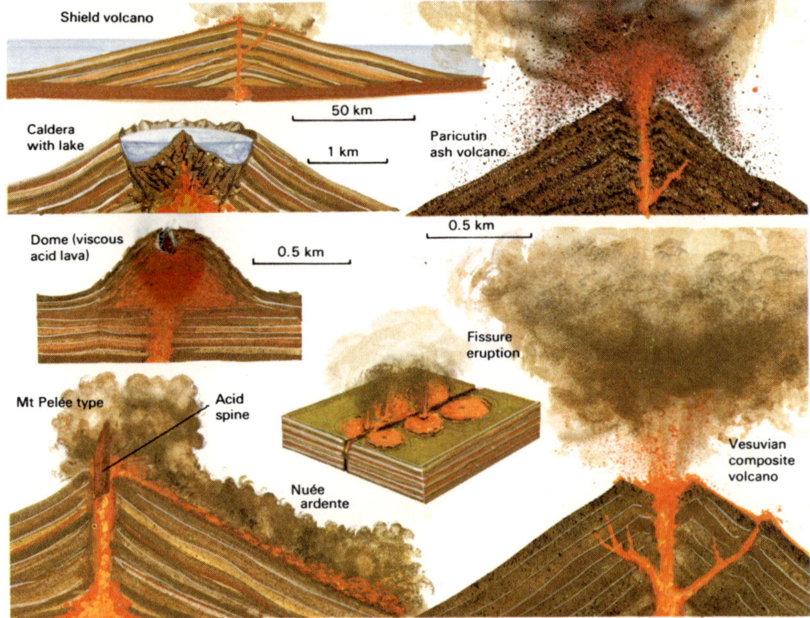

Right:
Typical examples of the main types of volcanoes.

First of all there is the type which is known as the *fissure eruption*. This is simple to describe. Here a flood of lava pours out of a crack in the Earth's crust and flows very freely. When it cools, it hardens into an almost flat sheet. This type of eruption is also known as the Icelandic type, because of the type of lava flows commonly found there.

In the Hawaiian type of volcano, so called for obvious reasons, lava pours out of a pit-like crater, and gas is quietly released most of the time. Occasionally, however, there is a sudden spurt of the volcanic gases blowing out a spray of glowing, burning droplets of lava. The well-known *Pele's hair* is caused by these droplets being caught in the wind and stretched into threads.

If the lava cannot flow easily (and this occurs when the composition of the magma is richer in silica), the gases have more difficulty in bubbling off. They are only released when the pressure builds up enough to force the gases out. In this kind of eruption, which is known as the *Strombolian type* after

Right:
A vent of a volcano may become plugged with solidified lava. When the cone is eroded away the plug or neck remains, like this one at Le Rocher-St-Michel in France.

Stromboli, Sicily, the volcano erupts from time to time carrying lumps of lava or *volcanic bombs*. Sometimes there are also lava flows.

If the lava is very thick and almost solid, great gas pressure is needed before it is released. When it does erupt it does so with explosive and terrible force, so that hot ash and fragments of lava are thrown high into the air. This is known as the *Vulcanian type*, and the characteristic cone is developed. There are other types of central volcanic types, that is, where the eruption is from one main centre.

Vulcanology is a very complex study, and, for the more adventurous, can be an extremely dangerous one, involving climbing to the very top of an active volcano. Under these conditions, even if the volcano is not actually erupting the ground may be so hot that it burns the shoes of the scientists and the gases may be almost suffocating. Workers have actually looked into open craters into the bubbling, burning lava below.

What comes out of volcanoes?

We have looked at the ways in which volcanic eruptions are caused, and at some of the many different types of volcanoes. Perhaps now, some of the products of volcanic activity merit further thought. We have already explained that the driving force behind many volcanoes is gas, but what kinds of gases are there, and where do they come from?

The most common volcanic gas is water, in the form of steam. If the eruption takes place where there is already plenty of water, such as in crater lakes or in the sea, for example, much of the steam results from water being heated. Some water is *juvenile*, however. Some of the water associated with the hot springs in the Yellowstone National Park in North America is juvenile, coming from the depths of the Earth. Much of it, however, is simply ground water that has been heated by contact with other volcanic rocks.

The other gases which have been found to come from volcanic eruptions are carbon dioxide, nitrogen, sulphur dioxide (it is this gas which gives the volcano its sulphurous smell), carbon monoxide, sulphur, and chlorine. They have been mentioned here in their order of abundance, that is, carbon dioxide is the commonest and chlorine the least common.

If you were asked to say what comes from a volcano, you would probably answer lava. Lava is the molten or semi-molten rock that pours from a vent during an eruption. It is usually at a temperature of between 900°C and 1200°C –

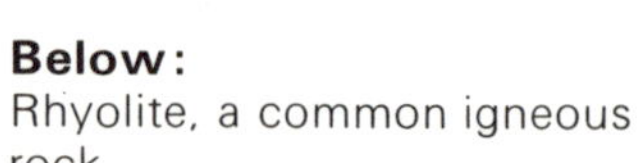
Below:
Rhyolite, a common igneous rock.

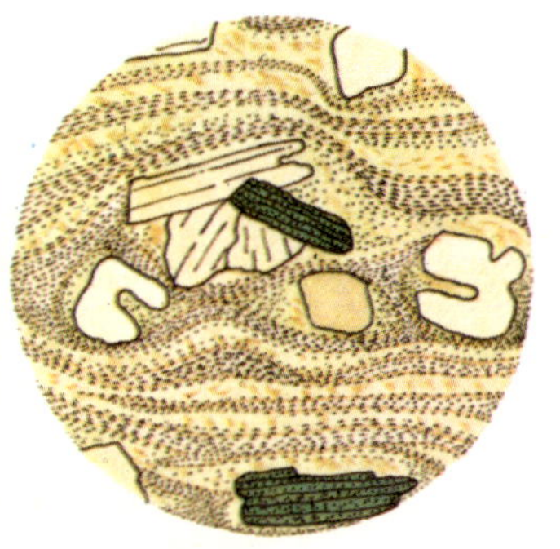

Rhyolite in hand specimen and through the microscope

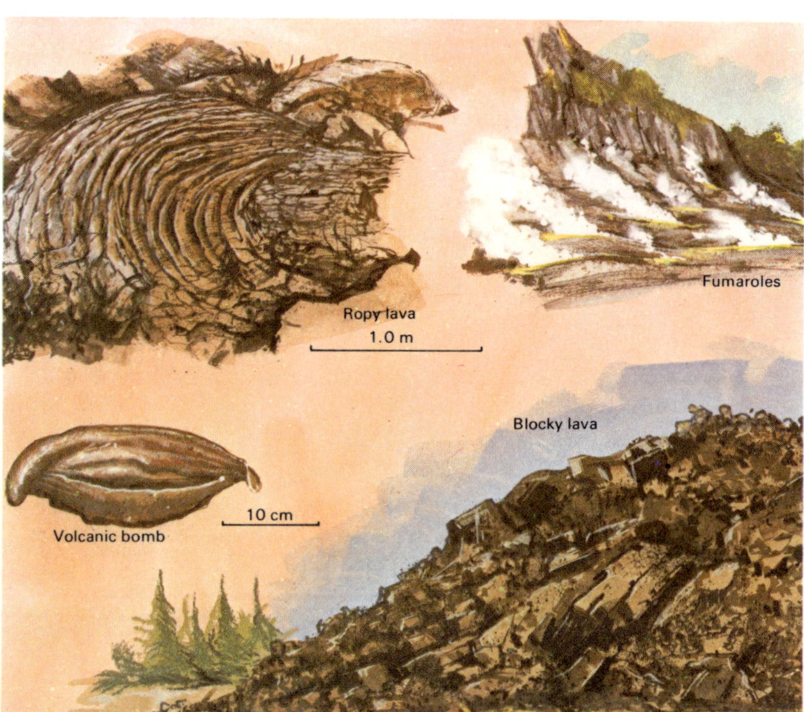

Right:
This illustration shows volcanic products including volcanic bombs and gas escapes called fumaroles.

Right and below: Some common volcanic rocks in hand specimen and under the microscope.

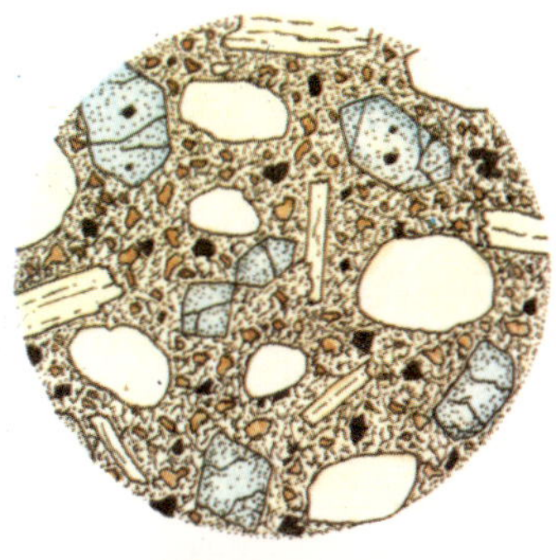

Basalt in hand specimen and through the microscope

remember how hot boiling water is and that water boils at a mere 100°C. This temperature depends upon the composition of the lava, but in any case it is never very much above the melting point of the rock.

Lavas can be acid (rich in quartz), basic (poor in quartz), or intermediate in composition. When the acid lavas have the composition of granite, they are called *rhyolites*. The basic lavas are called *basalts*. Acid lavas are usually so thick and pasty that they scarcely flow at all, and deposits which were once thought to be rhyolite lava flows are actually a material known as *ignimbrite*. Ignimbrite means fiery cloud and this is a very good description because it erupts as a cloud of gas charged with droplets of lava of a rhyolitic make-up, so that it can travel much further than the lava alone.

When a lava flow cools and solidifies, it usually has one of two types of surface. It can be blocky, in which case it is given the Hawaiian name of *aa*, or it can be ropy and then it is called *pahoehoe*. When lavas are spewed out under the sea, they cool very rapidly and form a pile of pillow-like structures, appropriately called pillow lavas. Sometimes, the lava is so charged with gases that when it cools it has a sponge-like form, and then it is called pumice. Pumice is so light that it will float on water. It is also the rock that you might have in your bathroom to remove stains from your fingers.

The other main type of volcanic products are collectively called *pyroclasts*. These are fragments of solid material that are thrown out of the volcano by hot, high-pressure gas. Volcanic bombs are typical of this type of material.

Where do you find volcanoes?

We expect you know where there are some active volcanoes either today or in the very recent past, because when one erupts it always makes the headlines in the newspapers. For example, we have already mentioned the eruption of Tristan da Cunha, and, of course, Mount Etna in Sicily is still active as are some of the volcanoes on the Hawaiian islands. Perhaps the most spectacular of the recent eruptions took place off the coast of Iceland when suddenly a new island, Surtsey, was born of the eruption. In fact, there are almost 800 volcanoes which are either active today or have been noted as having erupted during recorded history. Are all these volcanoes dotted about the surface of the Earth at random or can we see any sort of pattern in their distribution? And if there is a pattern, is it significant?

The answer to both of these questions seems to be yes. If we look at a map of the world which shows all the active volcanoes, we would see that about two-thirds of them are situated around the edges of the Pacific Ocean including many on the groups of islands. This grouping of the volcanoes is often referred to as the 'Pacific ring of fire'. Of the remaining volcanoes, a number can be seen ranging along the centre of the Atlantic ocean. It has been shown that this line of active volcanoes follows what is known as the mid-Atlantic ridge. This ridge is extremely important as we shall see later, but for the moment let it remain as a line of volcanic activity. There are two other areas where there are concentrations of volcanoes. In and around the Mediterranean area includes Etna and Vesuvius. The other area is east Africa where Mount Kilimanjaro is a good example.

It is also very important to note that although there may

Right:
Volcanic activity usually occurs at the margins of crustal plates, in particular, at those margins where crust is being created or destroyed.

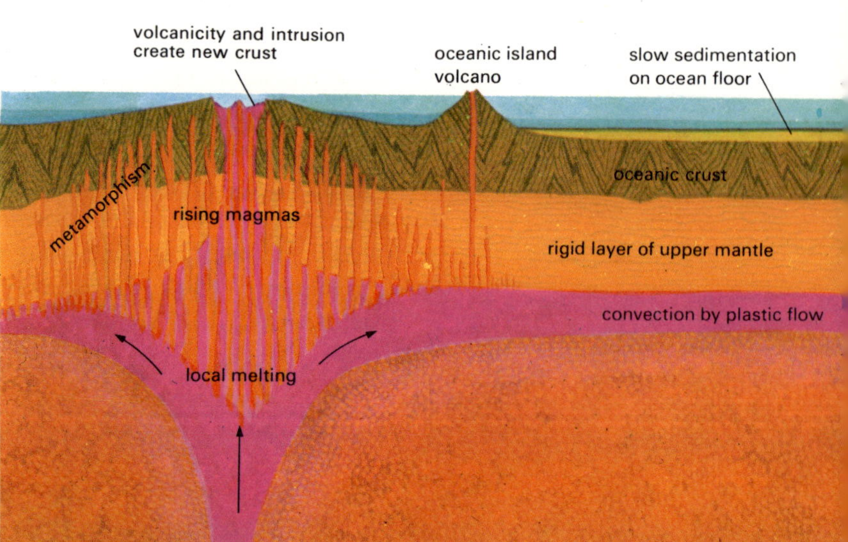

Left:
This map shows the world distribution of active volcanoes.

be slight variations from volcano to volcano in any one of these groups, each group tends to erupt similar products with similar compositions. For example, the mid-oceanic ridge volcanoes erupt basalt lavas while volcanoes in the 'ring of fire' often erupt lavas of andesite which contains more silica.

The significance of these patterns of volcanoes is concerned with the drifting of the continents but we shall deal with this more fully in another question. Suffice to say that the Earth's crust is made up of a number of rigid plates 'floating' on the mantle below. Concentrations of volcanoes are usually associated with the margins of these plates.

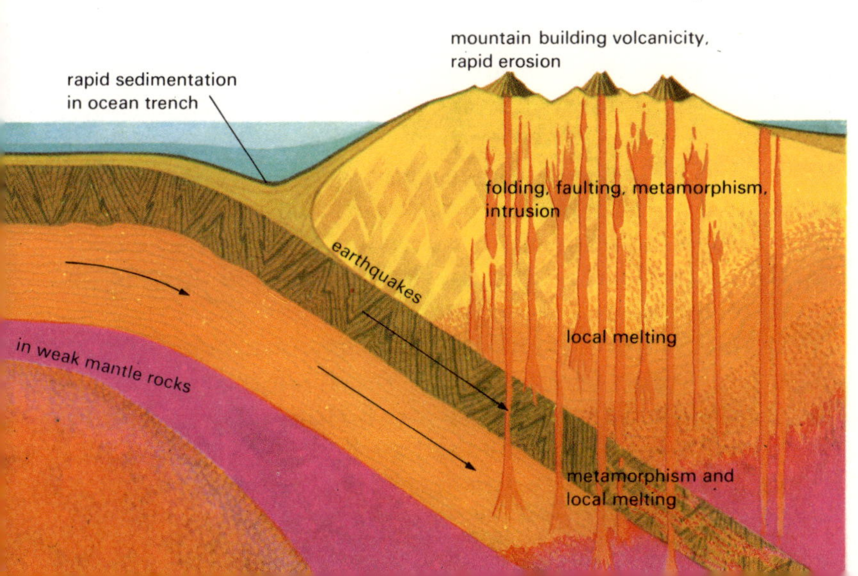

What are intrusions?

Igneous rocks are often divided into two major groups. These are *extrusive* and *intrusive* igneous rocks. We have already discussed the extrusives; these are the products of volcanic eruptions when magma reaches the Earth's surface in its molten state as lava and other fiery materials. Sometimes, however, the magma which moves into the crust of the Earth from the mantle below does not reach the surface. Instead, the magma cools and consolidates deep in the crust. Very often the rocks of the crustal surface lying over these blobs of rock are removed by the processes of weathering and erosion, so that the cooled and hardened magma is exposed to view. This type of igneous rock is called intrusive.

Right:
Various igneous intrusions.

Below:
Gabbro is a typical intrusive rock. It is shown in hand specimen and viewed through a microscope.

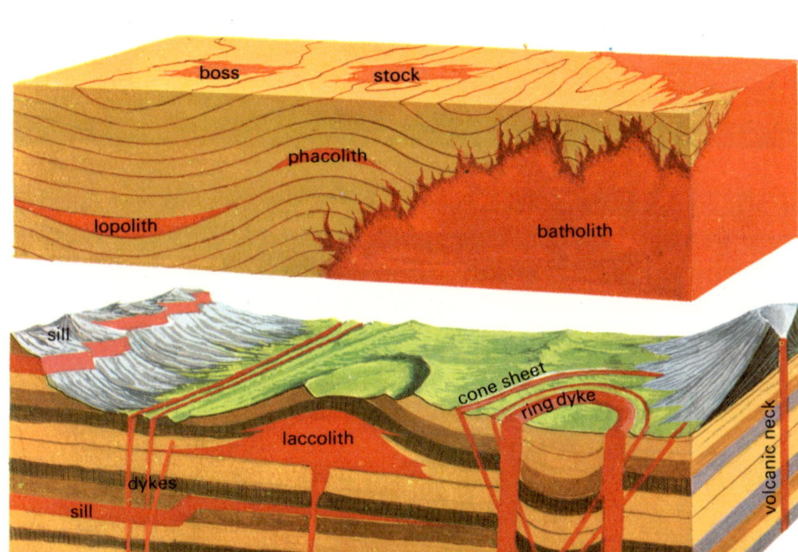

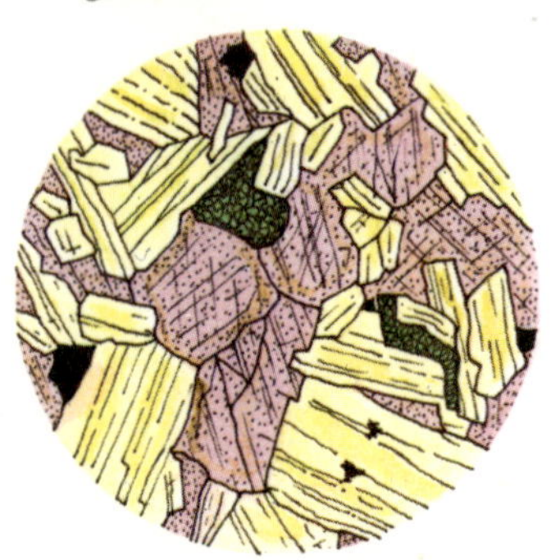

gabbro in hand specimen and thin section

There are many different kinds of intrusive bodies, depending upon the composition of the original magma and the conditions under which it is emplaced. In fact, the shapes of igneous intrusives depend to a large extent upon the relationship of the magma with the rocks into which it has intruded. These rocks are often referred to as the country rocks, and you should remember that they will be greatly affected by the heat and pressure exerted upon them by the intrusion. But more of that later.

Perhaps the best known intrusive igneous rocks are *dykes* and *sills*. Dykes occur when the magma has forced its way into the country rock cutting across any bedding planes. Sills, on the other hand, have managed to find the paths of least

Right and below:
Here are some more typical igneous rocks.

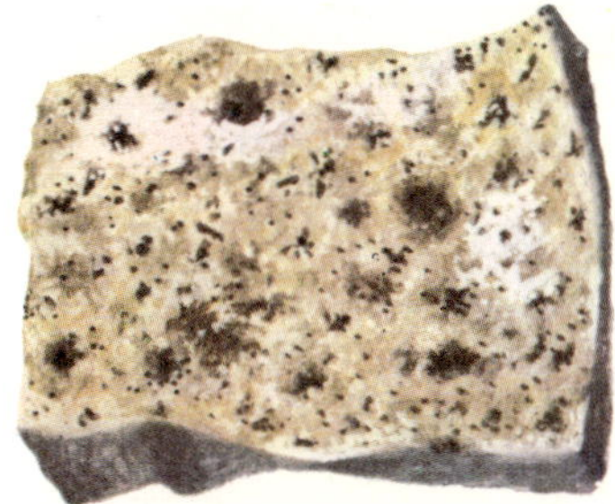

granite in hand specimen and thin section

resistance by moving along existing bedding planes. Sometimes, it is quite difficult to discover whether a sheet of appropriate igneous rock is a sill or a lava flow, that is, an intrusion or an extrusion.

The greatest and most complicated of all the world's intrusions are the *batholiths*. In Britain, perhaps the best known examples of granite batholiths form the cores of the wild and barren moors of the south-west of England; Dartmoor and Bodmin Moor, and also the areas of St Austell, Carmenellis, Land's End, and the Scilly Isles. In fact, tests have indicated that rather than being a number of separate small batholiths, all these granites are linked at great depths to form one giant intrusion. It is difficult to understand how igneous bodies of this immense size were able to force their way into the crustal rocks in the first place.

There seem to be at least two types of granite. One type may have been emplaced by 'floating' upwards while the roof rocks extended and then slid downwards. The other may have resulted by the conversion of existing crustal rocks by hot fluids.

What happens when a rock melt cools?

When a mass of igneous magma cools, there is a very definite series of events which take place. We begin on the assumption that the melt consists of various silicates; that is, compounds which are made up of silicon and oxygen atoms combined with certain metals such as iron, magnesium, and aluminium. This melt is probably nearer to being a solid than a liquid and is certainly under very high pressure conditions. The actual temperature at which the magma melts may vary from magma to magma, but it is always very thick and pasty. Petrologists describe this condition by saying that the magmas are very *viscous*.

You know that treacle, for example, does not flow as easily as water; therefore, treacle is more viscous than water. Actually, the melt is much less viscous if the magma contains

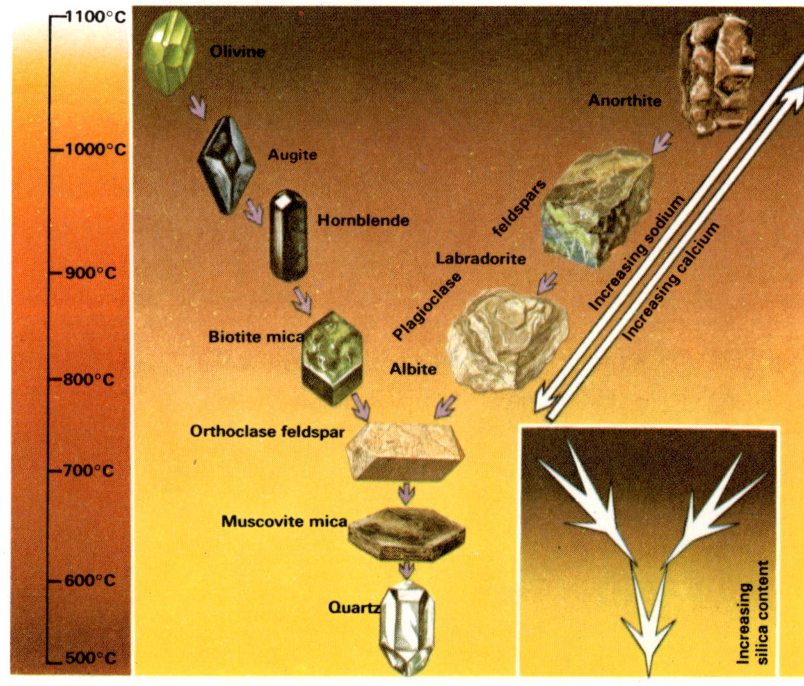

Right:
This diagram shows the order and the temperature at which the main silicate minerals crystallize from a magma. The cooler the temperature the higher the amount of silica.

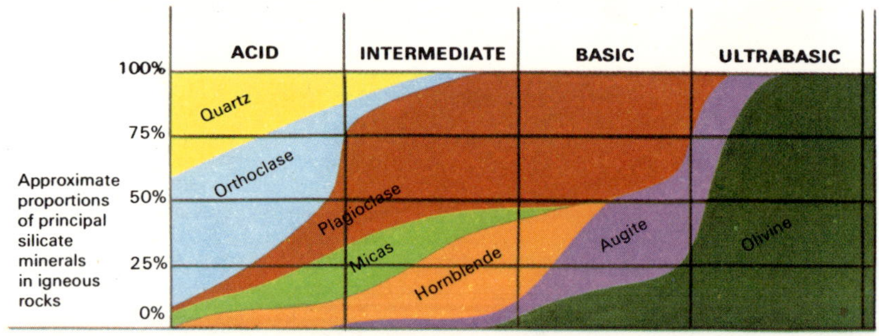

Right:
Igneous rocks may be divided into acid, intermediate, basic and ultrabasic groups. This diagram shows the amounts of each of the main silicate minerals in these groups.

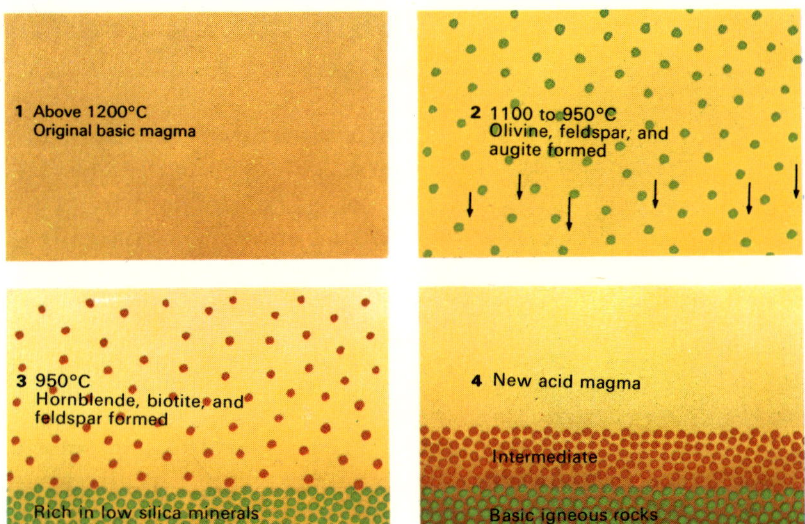

Right:
A single magma may produce different igneous rocks. As the magma cools, the first minerals to appear, sink because they are denser than the remaining magma. This settling of the crystals may allow a magma to give rise to two or more igneous rocks.

gases such as water vapour, carbon dioxide, fluorine and chlorine, and sulphur vapour. You can see then, that magmas containing between two and five per cent of these gases will flow more easily and will become solid at a lower temperature. A magma containing a lot of silica, such as a granite magma, will be at a much lower temperature than say a basalt magma.

As the magma cools, a temperature is eventually reached at which the first crystals begin to grow. Generally, a crystal requires some tiny particle around which to form, but it is not known exactly how this takes place in these conditions. The actual minerals that first form crystals will vary depending upon the composition of the original magma. It has been shown that when a crystal grows, the growth is faster on the edges and corners.

You may have already realized that rocks from volcanoes tend to be made up of tiny crystals whereas the granites, in the huge batholiths, for example, contain much larger crystals. This is because a giant mass of intrusive rock contains an enormous quantity of heat even though it may be at a lower temperature than the much smaller lava which contains less heat. The batholith cools in the crust whereas the lava cools in air. This means that a lava cools much more quickly than an intrusion. More crystals are able to form in the lava but they interfere with one another so that they cannot grow so large. This gives us one way of distinguishing a lava from a sill, a problem that we have already mentioned. Remember that the lava's upper surface will have cooled in contact with the air and will, therefore, have a fine-grained, chilled margin.

Which rocks are formed by water and wind?

Shale

Sandstone

Sandstone

Above:
Some typical sedimentary rocks.

Do you remember the Neptunists? They were a group of pioneer geologists who believed that granites for example, must have been formed by crystallizing from sea-water. Of course, we know now that this is not so, but there are rocks that do form in the sea and in lakes and rivers, too. These are the sedimentary rocks; there are many different kinds, some of which we have mentioned before. Do you remember what happened to existing rocks that were exposed to the power of the wind and rain? They were worn away and the rock fragments from them were carried off by the rivers and streams or in deserts by the wind.

This brings us to a rock that we expect everyone has heard of – *sandstone*. This is exactly what it says; a stone made of grains of sand. The word sand actually suggests a particular size of sedimentary particle but this need not concern us. Rivers and streams carry the mud, sand, and larger

Right:
Oolite, a type of limestone, is currently being deposited on the Bahamas Bank.

Right:
Sandstones are being formed as dunes in California.

Above:
More typical sedimentary rocks.

stones downstream until, eventually, it is carrying so much material and is flowing so much more slowly as it approaches the sea that some of the load settles. Naturally the larger boulders settle first, then the coarse sand, then fine sand and so on. The sea beating against the shore also removes material. All this sediment finds its way into the sea to be distributed on the sea floor. As more and more sediment is deposited, the lower level becomes harder and may eventually be compacted and 'glued' together enough to be called a rock. When the sea retreats and the area is uplifted by Earth movements we see the familiar sandstone.

If you have been to a beach where there is sand and pebbles you will have noticed that there are all sizes of particles mixed up together. When this hardens it forms another rock known as a *conglomerate*. Some famous examples of sandstones and conglomerates can be seen exposed in the cliffs of the south-east coast of Scotland. Some sandstones are formed by the wind, but these have different characteristics as we shall see later.

The other well-known group of sedimentary rocks is the *limestones*. There are many different types. The white cliffs of Dover are famous throughout the world, and they are made of one type of limestone called *chalk*. It is made up of millions of tiny skeletons of sea animals. The particles are usually called *coccoliths*. Limestones may also be formed by crystallizing from water rich in calcium carbonate. These are forming today in the warm seas of the Bahamas.

How can you tell if rocks are upside down?

Sedimentary rocks that are laid down in the sea, for example, are deposited in successive sheets, one on top of the other. As the conditions of sedimentation change, different types of sediments are deposited. For example, a series of sediments might be sands, muds, and then fragments of marine animal shells. When these were hardened, they would give sandstones, shales or clays, and then a limestone in that order. The geologist is particularly concerned with the order in which beds of rock have been laid down when he/she examines them in a cliff section, for example, so that the relative ages can be determined with the help of any fossils that might be present. But, as you know, Earth movements can completely alter the picture by folding the rocks or even turning them completely upside down. In our series of rocks then, how does the geologist work out whether it was the sand or the chalk that was deposited first? He/she uses what are often loosely called 'right way up structures'.

Try an experiment. Take a jam jar of water and drop a handful of sand and gravel into it. It is important that there are different sized particles. You should see that the largest fragments of gravel immediately fall to the bottom, the sand will be next, but the water will probably remain cloudy for some time because the smallest particles remain suspended in the water. This happens in the sea too. If you see a rock,

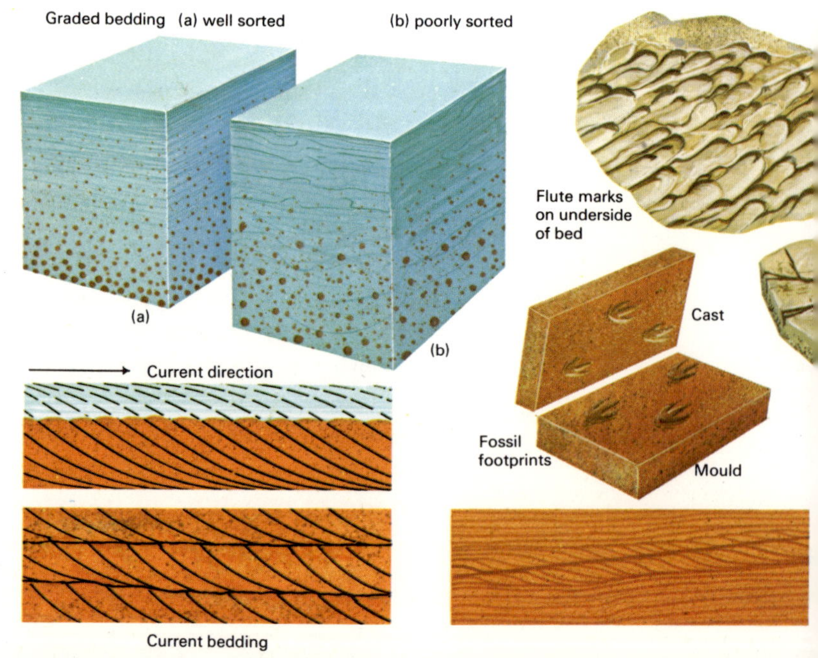

Right:
Some structures that may be formed during the deposition of sedimentary rocks. Some of these structures may be used to work out the order in which beds of rock have been laid down. The inset shows how the outcrops of rock could be interpreted in two different ways – the correct structure could be established with the help of sedimentary structures.

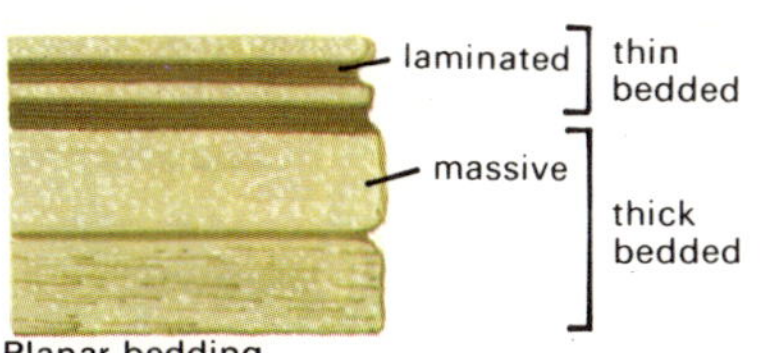

Planar bedding

Cross bedding in ancient sand dunes

Structures on the bottom of a bed of sandstone

then, which has larger grains at one side of the bed and finer grains at the other, the rock was originally deposited with the coarse grains at the bottom. This is called *graded bedding*.

We expect you have seen a dried out pond and noticed that there are cracks in the dried mud. These mud cracks can also be found in rocks, and, of course, they can only occur on the top of a bed, as can rainprints which are also to be found. Even the footprints of once-living animals can be preserved in the top of a bed of rock.

When sediments are laid down in water in which currents are flowing, the sediments show a pattern of fine bedding at an angle to the main bedding planes. This is called *current bedding*. If the top of the sediment is removed by erosion the tops of these fine beds are cut off. This is easily recognizable in the rock later. A similar situation also occurs when the beds have been laid down by the wind in a desert when sand dunes have formed. In this case it is known as *dune bedding*. These are just some of the tests that a geologist can apply.

Graded bedding

Cross bedding

Fossil ripple marks

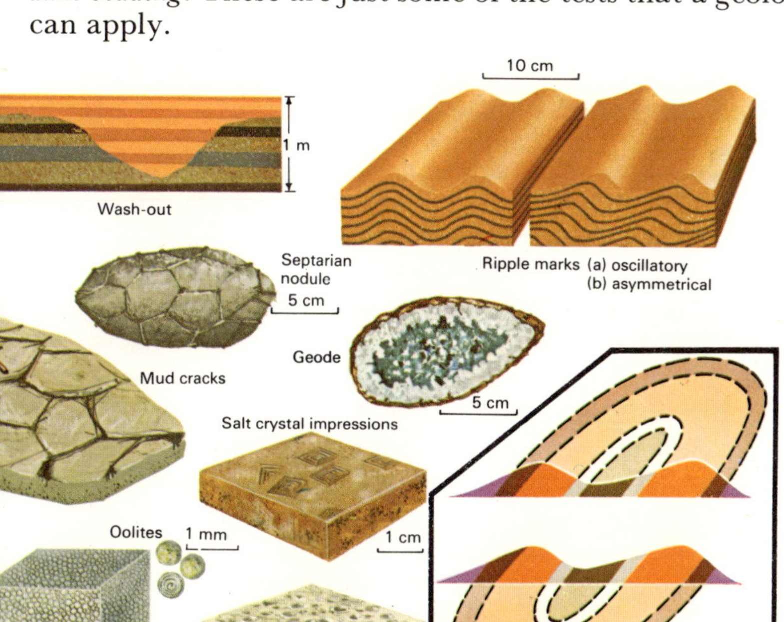

Above and above left: The way in which some of these structures illustrated on page 176 might appear in actual rocks.

What is an unconformity?

As has been mentioned earlier, sedimentary rocks may be laid down in the sea layer after layer, with the type of sediment reflecting the conditions of deposition at the time. Sometimes, however, the pattern is interrupted, and like all things on the Earth, change is taking place all the time.

Right:
Some different types of unconformity.

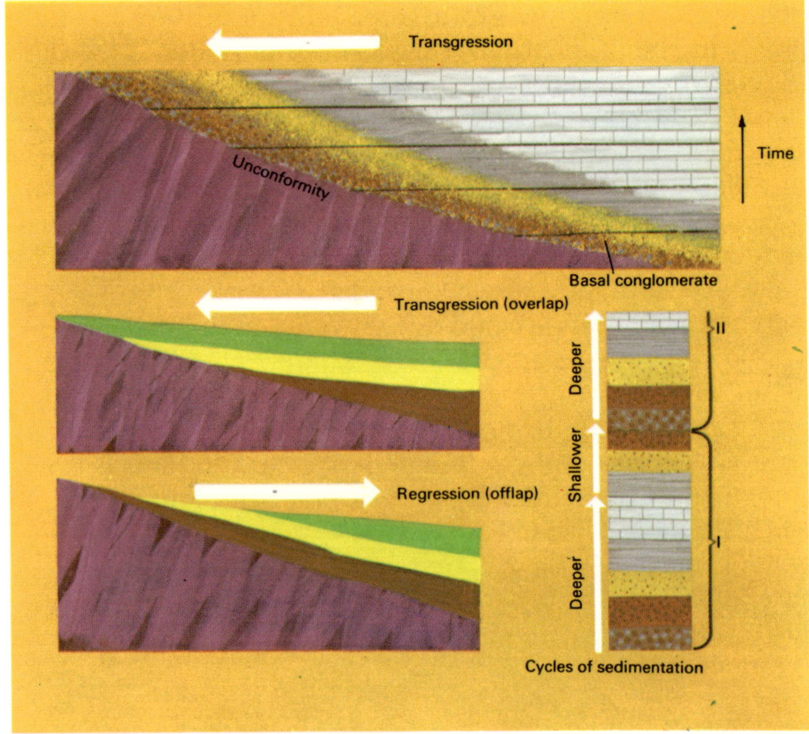

Try to imagine what would happen to the sedimentation as a result of a rising sea level, whether the rise be caused by a fall in the level of the land surface or an uplifting of the ocean floor. Clearly, as the sea level rose over thousands of years, the land would be slowly submerged. In turn, this would mean that deeper and deeper water sediments would be laid down further and further over what was once the beach and land.

As you can see on many beaches today, a common deposit is a mixture of sand and shingle. When this type of deposit has become hardened into a rock it is called a *basal conglomerate*. Thus, as the land submerges, the basal conglomerate covers a steadily widening area. Following the conglomerate, the rocks become deeper water in type and eventually *overlap*. The sea level may fall again, of course, and this is known as a marine *regression*. The rise is known as a *transgression*. When the sea level falls, obviously the sediments become shallower water in type and they tend to *offlap* as you can see in the picture.

But what of the rocks forming the beach itself? They may have been dry land for millions of years, and in that time, they have been folded and metamorphosed. Can you imagine what a section through this area might look like when it has all had time to become hard rock? First of all, at the bottom, you would see layers of old rocks folded, and on top of them you might see further layers of horizontal beds representing the rises and falls in the sea level. The boundary between the two sets of rocks is known as an *unconformity*. The important thing to remember is that an unconformity represents a break in time. In other words, there was a time gap between the laying down of the first rocks and the second. If this gap was long enough, it may have been that rocks of the older period were worn away by the elements of weathering and deposition into a typical up-hill and down-dale shape. In that case the boundary between the older and younger rocks will be an *erosion surface*. It may be that there is more than one unconformity in a section if the above process has occurred more than once.

Sometimes, the younger rocks may be lying exactly parallel to the older ones, and may even be folded into the same shape by a later period of folding. In this case, the unconformity simply shows that there was a break in deposition with neither folding nor erosion between.

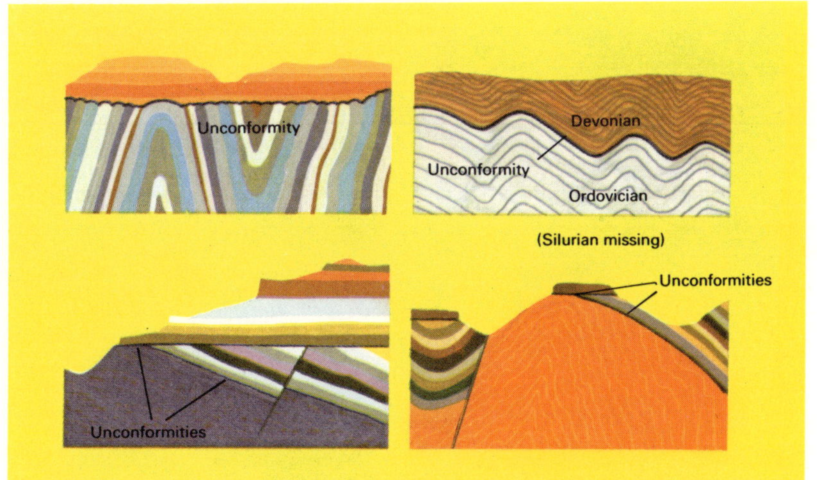

Right:
A series of transgressions and regressions like these might deposit sediments in a cyclic way.

How are rocks 'changed'?

You know now that there are three major groups of rocks and we have already looked a little more closely at two of them, the sedimentary and the igneous rocks. The other group consists of rocks to which the famous geologist, Sir Charles Lyell, gave the name of metamorphic rocks. These are the rocks that have been changed either by the heat of an igneous body, in which case it is *thermal* metamorphism, or on a much larger scale by heat and pressure within the Earth, when it is known as *regional* metamorphism. Metamorphism causes the affected rocks to form new minerals while the whole composition of the rock may or may not remain the same. The important point is that the rock does not pass through a liquid stage as the changes occur. You can get an idea of what is happening to these rocks if there is a fall of snow in your area during the winter. Take a handful of snow and squeeze it, and you should see that it changes from the fluffy crystals of snow into much harder, more compact crystals of ice. This is called *recrystallization*.

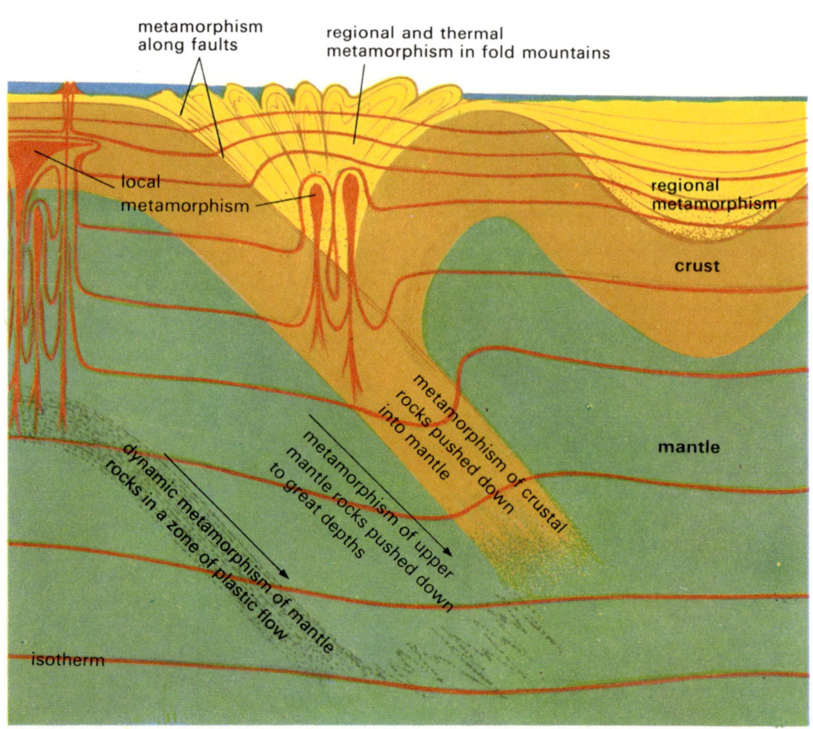

Right:
Some of the ways in which enough heat and pressure occurs for rocks to be metamorphosed.

All rocks can be changed by metamorphism, including those that have already been metamorphosed before. You have probably seen columns made of a stone called marble in your local museum or country house, or you may have

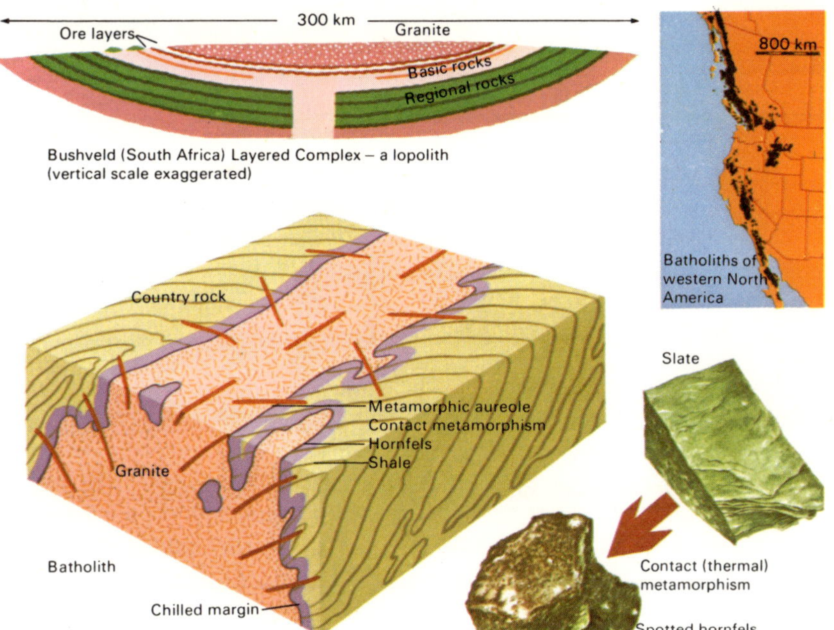

Right:
A large igneous intrusion may contain enough heat to form a wide band of thermally metamorphosed rocks around it even though the intrusion may not be as hot as a lava flow.

seen slabs of marble that have been cut and polished to form a table top. This is a metamorphic rock. To be exact, it is a metamorphosed limestone. You will remember that limestones are made up of calcium carbonate which may be formed by the shells of marine animals. If limestone is heated and subjected to pressure, all the shell fragments disappear and the calcium carbonate recrystallizes to form the beautifully patterned white marble that is found in such places as Carrara in Italy. Although some pressure is involved in this change, it is mostly brought about by heat, and is thus an example of thermal metamorphism.

Regional metamorphism tends to develop a distinctive family of rocks because in addition to the heat, there is immense pressure and movement which occurs in association with the Earth movements causing the metamorphism. A typical regional metamorphic rock is slate. Slate used to be widely quarried for roofing houses and as the bed of the best billiard tables. Can you think why? It is because slate can be split along very flat planes. These flat planes arise from the fact that all the minerals which have resulted from the recrystallization of the shale are lying along one direction, because of the pressure which has been applied.

How can you recognize metamorphic rocks?

Metamorphic rocks arise from all types of rocks and can be extremely variable, but they all have the fact in common that they have been modified by heat and/or pressure. You have seen the way in which a limestone may be baked to form a marble and how a shale or mudstone can be heated and pressed to give a slate. There are a number of features which will enable you to say that a rock is or is not metamorphic at a glance. Of course, you still need to be able to say which metamorphic rock you are looking at before you have made a really useful statement.

The first points which you might see are the textures and structures within the rock you are looking at. If the rock is a metamorphic rock, it will possess some characteristics which are purely a result of the metamorphism, and perhaps some which would have been found in the original rock. The way in which the minerals have grown as a response to the stress exerted upon the rock has a great influence on the appearance of the end product.

Right:
Some typical metamorphic rocks in hand specimen and viewed through the microscope. Notice the schistosity and banding that may often be seen in rocks that have been subject to heat and pressure.

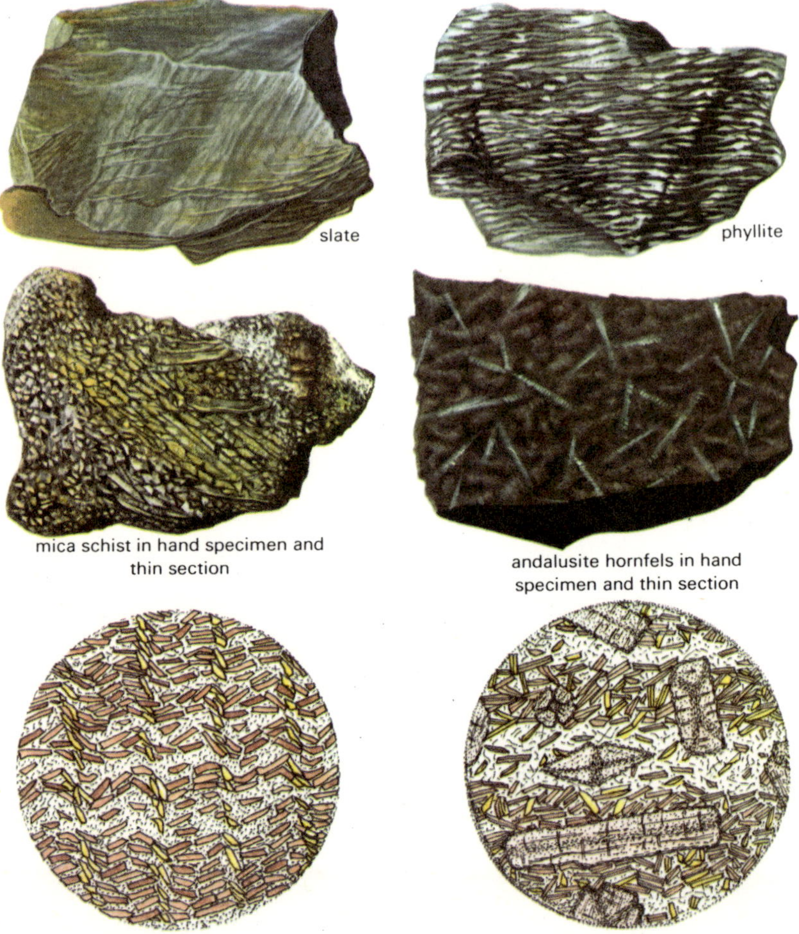

quartzite in hand specimen and thin section

impure marble with calc-silicates in hand specimen and thin section

Right and below: Here are some more typical metamorphic rocks.

gneiss

marble

Because of the distortion that has taken place, the minerals in a metamorphic rock do not often have the good crystal outlines which you would expect to find in a granite. Similarly, minerals such as mica which often occur in plate-shaped crystals, develop this tendency to an extreme and are often lined up in a particular direction so that the rock may have a flakey, crumbly appearance often glistening with the flakes of mica. This appearance is usually called *schistosity*. You may also find that if there are any larger grains of minerals in the rock, these large grains may contain fragments of other minerals included within them. This results from the new minerals growing around unchanged grains.

Another common occurrence which is well illustrated in the metamorphic rocks of the Pyrenees Mountains is that some minerals, like garnets, may show indications of having been rotated during the metamorphism. This indicates that the rocks have been subjected to more than one period of metamorphism. The garnets have formed as regular crystals during the first metamorphic episode, and then during the second have been partially remelted and then turned.

You may sometimes find that metamorphic rocks have a distinctly banded look, in which light and dark bands of minerals often alternate. If the original sedimentary rock was banded, the banding may have been preserved during metamorphism. This texture is often seen in metamorphic rocks known as *gneisses* (pronounced nices). You must be careful to look for those characters which are purely metamorphic and those which are original.

How can you tell the exact age of rocks?

The answer to this question is perhaps a little more difficult to understand than some of the others you have looked at. But if you read through the explanation slowly, think hard and look carefully at the pictures you should soon be able to work out the method. Once you have grasped it, you will never forget it, and it will be a very useful addition to your mine of knowledge.

The most accurate way in which rocks, fossils, ancient relics, or even the Earth itself can be dated is by a process which is usually called *radiometric dating*. It depends upon a property that was noticed in naturally occurring elements such as radium, which was first discovered by the Curies in 1898. This property is radioactive decay.

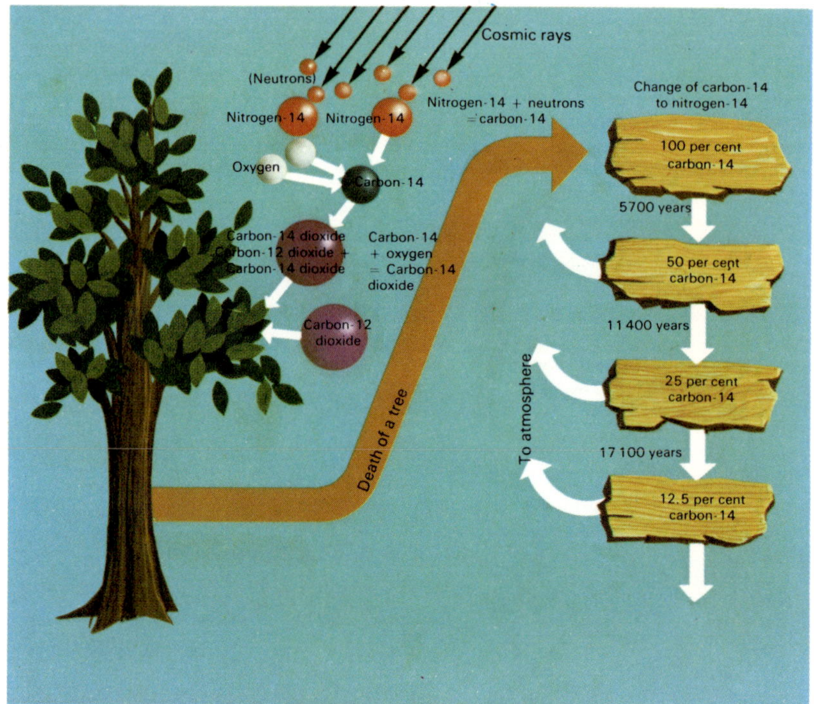

Right:
Cosmic rays convert nitrogen in the air to an unstable isotope of carbon – carbon 14. This in turn combines with oxygen from the air to form carbon dioxide. Plants take this in together with ordinary carbon dioxide. When the plant dies the carbon 14 reverts to nitrogen at a known rate so that the age of a dead tree can be worked out.

You know that elements are made up of tiny particles called atoms. Now, atoms are composed of even smaller particles called *protons* and *neutrons* in the centre, surrounded by what can be imagined as a cloud of electrical charge called *electrons*. An atom of a particular element must always, by definition, have the same number of protons and electrons, but sometimes the number of neutrons may vary. When this occurs the resulting atom is an *isotope*. For example, an atom of carbon may normally contain twelve

Right:
This device is used for accelerating atomic particles. It is called a cyclotron.

Below:
The half-lives of some radioactive elements that may be used to date rocks.

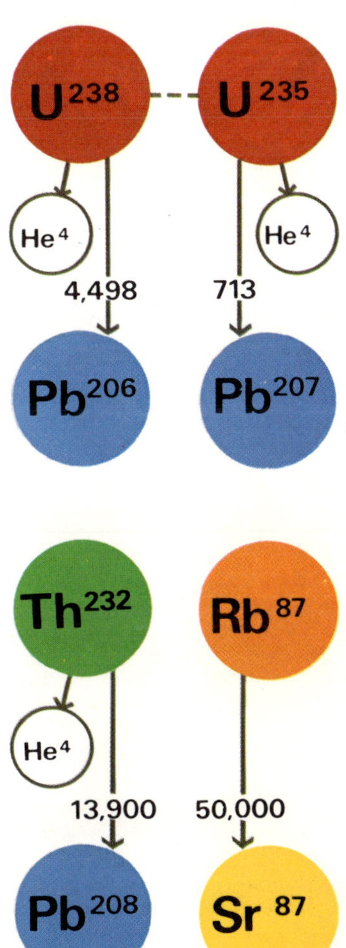

neutrons but it can contain fourteen. Isotopes are peculiar, in that they may change into other atoms by losing particles quite naturally. This takes place at a constant rate. The rate can be measured by the *half-life* – this is the time it takes for one half of a given weight of a radioactive element to change into its daughter elements, as they are usually known. The half-life is constant for any one element.

The radioactive isotope of the metal uranium containing 238 neutrons instead of the stable 235, changes into lead containing 206 neutrons and the gas helium containing 4 neutrons. The half-life of this chain is 4498 million years.

Now when igneous rocks crystallize, for example, we have an exact point for the beginning of the radioactive decay of any radioactive material contained in it. To work out how long ago the rocks were formed, it is only necessary to work out the amount of original, parent, element there is compared to the amount of daughter element. This sounds simple, but in practice requires a great deal of skill and knowledge. This technique has meant, however, that famous geologists like the late Arthur Holmes were able to date the Earth more accurately than ever before at about 4500 million years old and to place the base of the Cambrian system at about 600 million years instead of the 500 million year date that had been established by traditional methods.

How should you collect rocks and minerals?

The basic equipment required for collecting rocks and minerals is the standard geological gear of hammer, chisel, notebook and waterproof pen or pencil, newspaper for wrapping, and something to carry home your finds.

There are one or two other items that are particularly useful for the rock hound. If you are collecting rocks, it will not only be the soft easily broken types that you will wish to take home. You will probably want samples of igneous and metamorphic rocks which may be very hard. It would be difficult to break off pieces of these rocks with a normal geological hammer. A valuable item of equipment in a hard-rock area is a sledge hammer. But remember that this type of hammer may weigh more than four kilos, so that it will be very heavy to carry and could be extremely dangerous – missing the rock could easily result in a broken leg. If you wish to take advantage of this hammer, make sure that you are accompanied by someone who knows how to handle it properly. When hammering on hard rocks, splinters of rock may sometimes fly off and could easily hit you in the eye. Again in hard-rock areas, you would be well advised to wear protective goggles over your eyes. Small specimens of delicate minerals should be carefully wrapped and placed in a matchbox to prevent them from damage by rubbing against hard rocks in your bag. Do not forget that a hand lens is very useful for field identification of rocks and minerals.

Right:
An exhibition at the geological museum of the Institute of Geological Sciences, London.

Right:
A geologist at an outcrop in Norway.

Before you set off, consult a geological map and a regional guide, and pay a visit to your local museum. The best places to collect rocks and minerals are cliffs, quarries, mines and their spoil heaps, road and rail cuttings, or in more rocky terrain almost anywhere. Remember these are all dangerous areas, so make sure you are accompanied by an experienced person.

You know that rocks are simply aggregates of minerals so that where there are rocks there are minerals. The best crystals, however, can only be found in certain places. An ideal place to look for good mineral veins is where large bodies of granite have been intruded in the country rock. You can even find minerals on the beach, but they will have been worn by the action of the sea. Some beach deposits are almost entirely made up of one particular mineral such as agate or very rarely olivine.

Many people find the collection of fossils and minerals a fascinating hobby because of the beauty of the specimens or because they represent all that is left of once living creatures. Fewer people find the collection of rocks so exciting. Rocks, of course, are more difficult to identify without the aid of a thin section, but every rock, mineral, or fossil gives a clue to the complex history of the Earth.

Would a compass needle always have pointed in the same direction?

This seems to be a very strange question to ask doesn't it? Perhaps you will see that it is not as strange as all that. We expect that you have seen or even used a compass. If you haven't, it is simply made, and you could make one for yourself.

Take an ordinary darning needle and a bar magnet. Using one end of the magnet only, stroke the needle with it, making sure that you always make the strokes in the same direction. You must also make sure that as you come to the end of each stroke you bring the magnet back to the other end at some distance from the needle. After a little while, you should have made a weak bar magnet of the needle. Now place the needle on a piece of cork (a slice of ordinary bottle cork would do), and float the cork in a bowl of water. You should find that no matter which way you turn the bowl, the needle on the cork always moves so that it lies in the same direction. You have now made a simple compass, and one end of the needle should point to the south and one to the north. Of course, it is possible to buy much more accurate and expensive compasses.

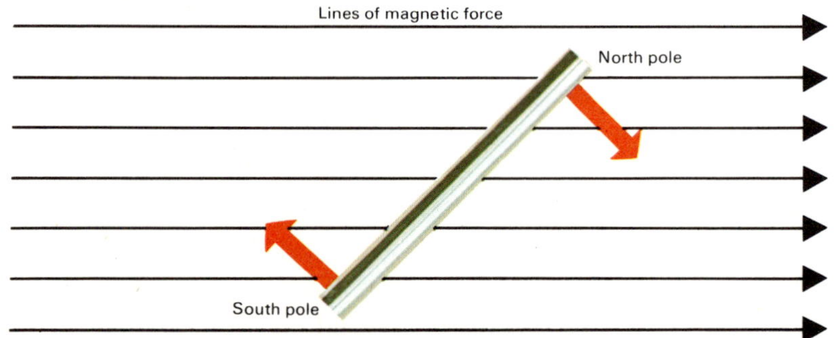

Right:
A bar magnet placed in a magnetic field will be forced to line up with the lines of force.

The reason for the almost magical behaviour of the compass is that the Earth has a *magnetic field*. Explorers have taken advantage of this for centuries in navigating the surface of the Earth. In fact, the magnetism of the Earth is rather like having a bar magnet placed along the middle of the Earth along a line between the north and south poles.

There have been a number of ideas about the origins of the Earth's magnetism. For example, the Earth may be working rather like the dynamo on your bicycle only in reverse. That is, instead of the movement producing electricity, perhaps electricity is producing the magnetism? It

Right:
The lines of force associated with two bar magnets, with north and south poles adjacent.

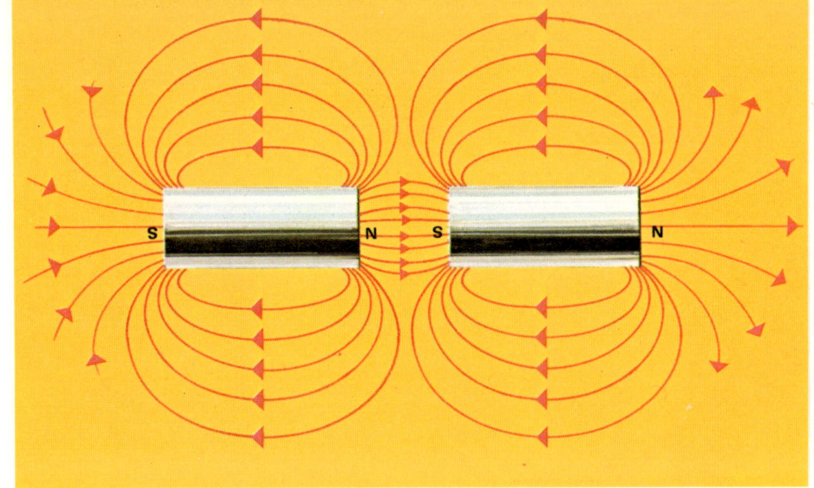

Right:
This is an auroral display. It is also known as the Northern Lights. It is the result of the effects of the Earth's magnetism on charged particles approaching the Earth. The effects are more marked during increased sunspot activity.

can also be shown that at the moment the magnetic poles are not on the Earth's axis of rotation.

You must remember that when igneous rocks cool, certain minerals such as magnetite actually take on a tiny magnetic field which corresponds to the magnetic field of the Earth at the time of cooling. It is possible to measure the direction of these tiny fields in ancient rocks by means of very sensitive instruments.

It can be shown quite definitely that not only do the poles of the Earth wander about their present position but they also reverse. The north pole becomes the south pole and vice versa. This means that at certain times during the Earth's history, a compass needle would have pointed in the opposite direction to the way in which it points today. It is hard to imagine, however, exactly what happens to the Earth when the change is taking place. The reversal seems to occur quite quickly so far as geological time is concerned.

What are mountains made of?

From what you have learned already, it is obvious that mountains must be made of rock. But what kinds of rocks and what kinds of mountains are there? How do the mountains get there in the first place? Over the years in which geological thinking has been at its most active, many theories have been put forward to answer these questions. In fact, these types of questions hold the key to some of the most important facts concerning the Earth's whole history. These are the ideas of mountain building.

If you were asked to name a mountain that you know, which one would it be? Perhaps it would be Mount Everest or Mount Etna or perhaps you might suggest the Harz mountains of Germany or the Juras of France and Switzerland. All these mountains are different; made of different types of rock and formed by different processes. Perhaps the simplest and easiest example to understand is the volcano, Mount Etna. Its cone is built up by successive outpourings of lava from the bowels of the Earth. The Harz mountains were formed by block faulting, in other words they are the *horsts* that we have already mentioned.

Right:
The work of the Earth's upheavals provides challenges to intrepid mountaineers.

Right:
A section across the ridge that is present in the middle of oceanic basins. At this point, new crust is being formed and it pushes the land masses further apart.

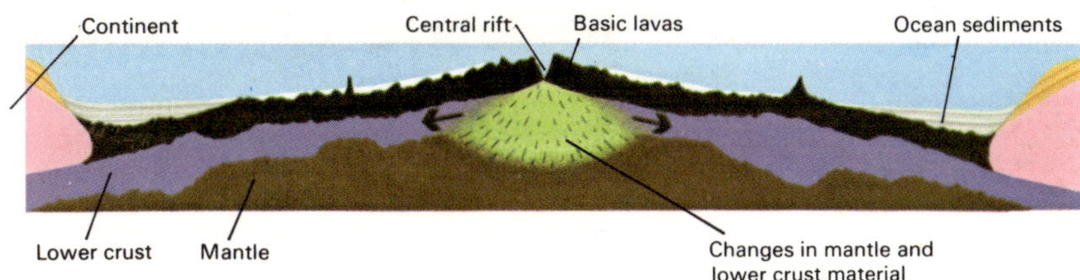

In these two examples, the volcanic mountain such as Etna can be made of a variety of volcanic materials such as basalt lava or cinders and ash, or more acid, blocky lavas, and the fault mountains can be uplifted blocks of granite such as those that form the Kharas mountains in south-west Africa.

You probably know that Mount Everest is the highest mountain in the world, standing 8848 metres above sea level. (It is interesting to note that the deepest part of the ocean known to man is the Marianas Trench south-east of Japan, which is 11,035 metres deep.)

You might also know that Mount Everest does not stand alone but is part of a great chain of the highest mountains anywhere, called the Himalayas, astride the borders of Tibet and Nepal. Among these lofty pinnacles the Abominable Snowman, the *Yeti*, is said to have his icy home.

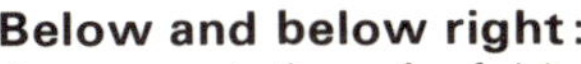

Below and below right:
A representation of a folded mountain chain and some typical rocks that might be found in it showing how they become more intensely metamorphosed towards the core.

1 Shale

2 Schist

3 Gneiss

4 Granite

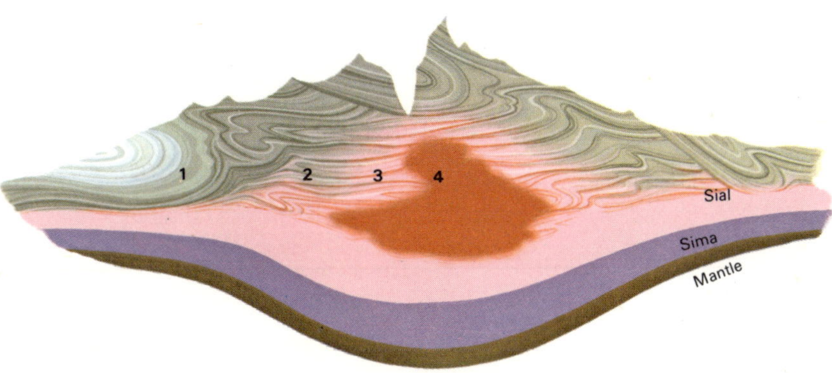

There are many other such chains of mountains; for example, the Alps or the more modest mountains of the Highlands of Scotland. These are the great *fold mountains*, made of sediments laid down in primordial oceans and then uplifted by crumpling of the Earth's crust during periods of crustal shortening. The fact that these mountains occur in long chains in certain areas of the globe is no accident as you will see later, but is associated with one of the most important concepts of the modern Earth sciences, that of continental drift, the almost unbelievable idea of whole continents moving.

How are mountain chains built?

Most of the major mountain chains of the world have been constructed by periods of accumulation of unusually thick deposits of sediments in elongated basins followed by uplift, folding, thrusting of huge masses of rock called *nappes*, and faulting. These processes, known collectively as *tectogenesis* are usually accompanied by intense volcanic activity as well as the forcing into position of huge masses of granite which may continue to rise for a long time after the main movements have ceased and the processes of weathering are active. These regions in which Earth movements are concentrated are usually given the name of *orogenic belts* and the whole process is known as an *orogeny*.

You can see that the process of orogeny involves the sinking of a part of the Earth's crust, followed by sedimentation in the basin that has been formed, and finally uplift of the whole area. It is interesting to note that mountain chains such as the Juras that we have already mentioned are formed by a different process. In this case, certain parts of the nearby Alps had been uplifted by orogeny so that the area now occupied by the Juras was tilted. It so happened that the rocks which now compose the Juras were laying upon a huge bed of salt. In effect this salt acted like a layer of lubricating oil so that the rocks on the top slid very slowly down hill folding and faulting as they went. This is known as *gravity tectonics*.

In 1859, the geologist, James Hall, working in the northern Appalachian mountains of America discovered that the sediments that made up this range were of a type that could only have been deposited in quite shallow water in the sea. At the same time he recorded that the sediments were in some places as much as 12,000 metres thick. How, then, did such a thick deposit accumulate? It is clear that the older

Below:
The theory of isostasy suggests that as mountains are worn away by erosion, the Earth compensates by further uplift.

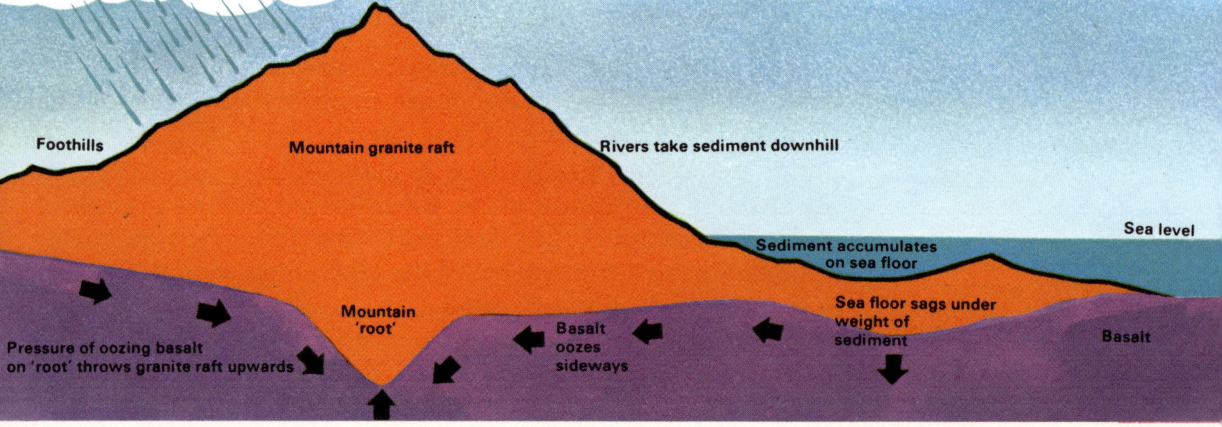

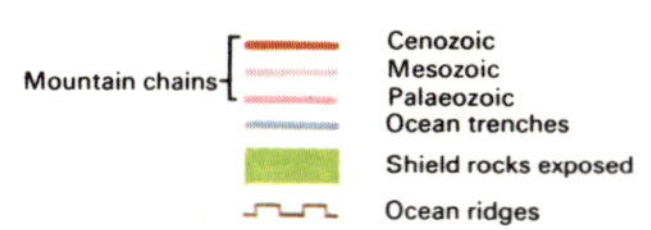

Right:
The positions and ages of the world's mountain chains.

rocks underneath must have been sinking as fast as the new sediments were being laid down. This type of elongated, sinking basin has been called a *geosyncline*. At first, it was thought that the sinking was caused by the weight of the sediments being deposited on top but it was soon realized that this was not so. In fact, it was the famous mineralogist, Dana, who first suggested in 1873 that the sinking of the basin was the cause of and not the result of the sedimentation. Nowadays a new, all-embracing theory seems to provide us with all the answers to the problems of how and why mountain building takes place in the way it does. This is the theory of *plate tectonics*.

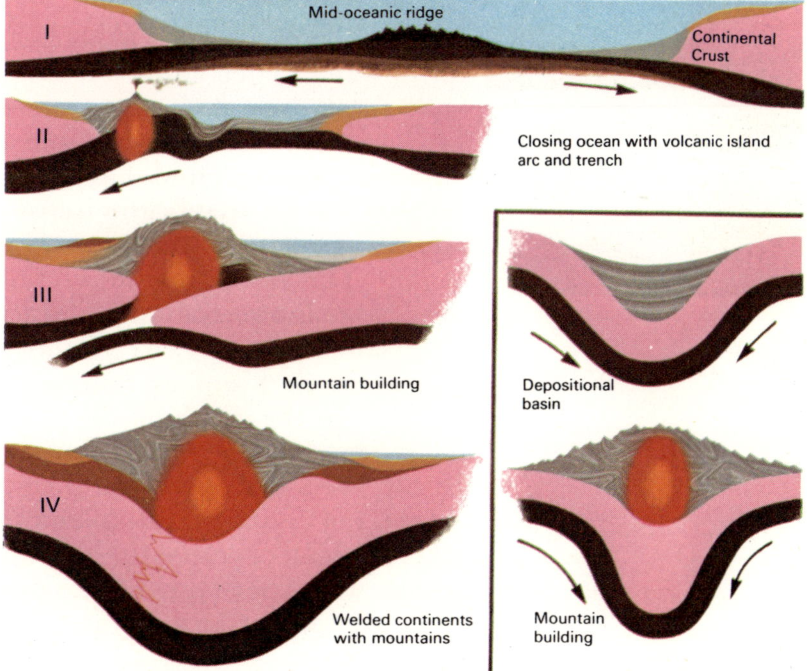

Right:
The all-embracing theory of plate tectonics neatly accounts for the method of formation of mountain chains. The inset shows the more traditional theory for mountain building with the crust being dragged downwards, filled with sediment at a rapid rate, and then squeezed.

Why do mountains form?

You have already seen what some of the world's mountains are made of and the way in which the fold mountain chains are formed by crustal sinking, sedimentation, folding, and uplift. You should also note that many of the rocks are subjected to intense pressures and heat so that they are

Right:
A section through the Earth's crust showing the continent and ocean basin. The arrows indicate major movements.

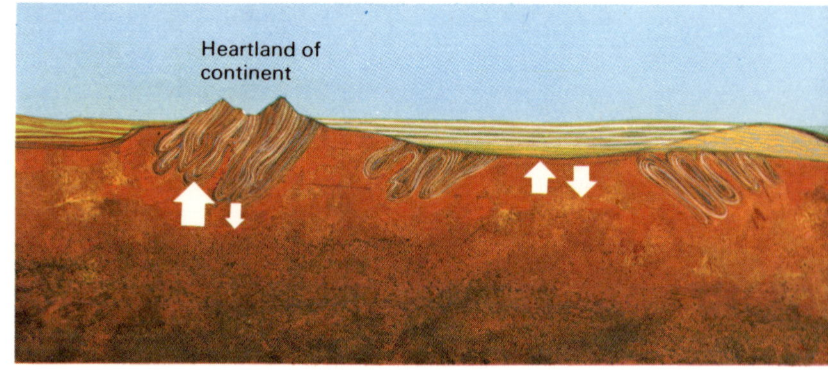

Above:
A typical mountain scene.

metamorphosed to the schists and gneisses characteristic of these areas. We have stated that the crustal sinking causes the sedimentation but why should it occur at all?

You know that beneath the Earth's crust and before the core is reached, the mantle of the Earth is intensely hot and has a semi-plastic consistency. It has been suggested that within this zone the material is slowly circulating in a number of what are usually referred to as *convection cells*. You can get an idea of what a convection cell might look like by means of a simple experiment. Take a heat-proof glass dish and fill it with water. Then add a few drops of ink or some other dye, and slowly heat the water preferably using a Bunsen burner or a candle with a single flame at the centre of the base of the dish. You will find that as the water closest to the source of heat warms up it begins to rise as shown by the ink. It will slowly rise to the top where it will cool and begin to fall again. This is a simple convection cell. Some of the decorative table lamps containing different coloured oils work on the same principle. It is thought that mantle material circulates in a similar fashion – but much more slowly, of course – in the depths of the planet.

Where two downward flowing currents are side by side it has been suggested that their movement drags the crust of the Earth down as well, and that this is the beginning of the orogenic cycle with the formation of the mountain roots. (It has been proved that mountains do have roots; that they are rather like icebergs, but instead of floating on water they rest on the dense, viscous mantle.) As the currents develop and move more rapidly, the downward drag is at its maximum with rapid sedimentation and metamorphism in the core. Finally, it has been thought that the currents begin to die and the already formed mountains are slowly uplifted rather like a cork rising from the bottom of a dish of water as soon as it is released, but again in the case of the Earth

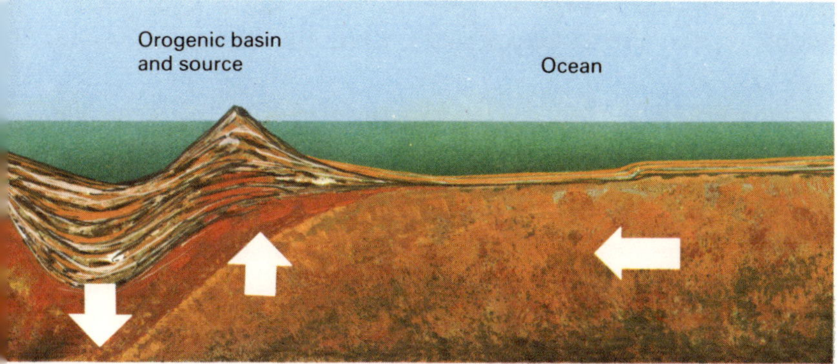

everything happens more slowly. Unfortunately, this simple theory is rather too simple to account for the complex happenings of mountain formation, but once again, as we shall see the theories of plate tectonics come to the rescue.

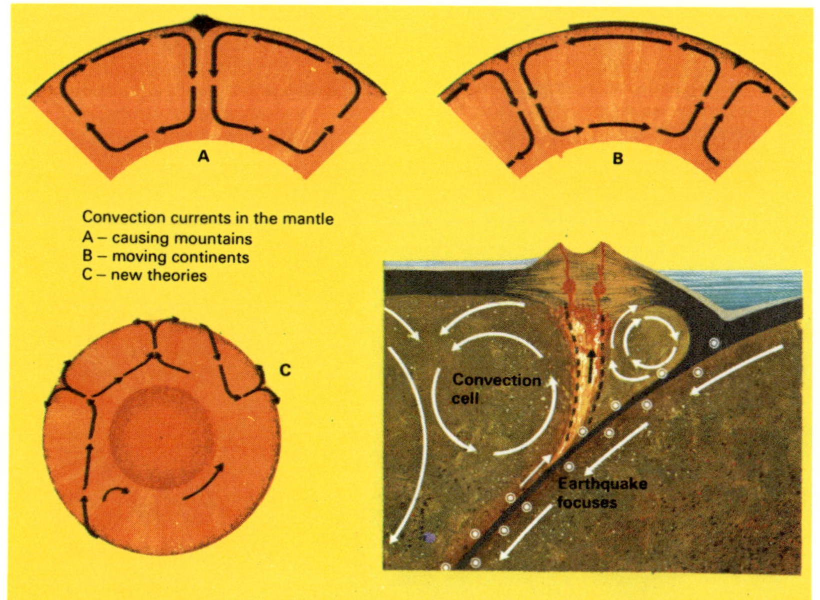

Left:
A representation of possible convection currents in the mantle. You can see how new crust can be formed at the mid-oceanic ridge – the result of diverging convection cells.

Can whole continents move?

It is almost impossible to believe, but without doubt, the answer to this question seems to be yes. But how do we know, because we cannot see or feel them drifting beneath our feet? In the early days of geology, the drifting of whole continents was certainly not considered to be a possibility, and in fact, in 1846 Dana positively asserted that continents and oceans have always been in the same place throughout geologic time. What happened, then, to change the geological mind?

It was a man called F. B. Taylor who first put forward a convincing argument for the movement of whole continents. He made this statement in the year 1910. During the following two years H. B. Baker and perhaps most famous of all A. Wegener supported Taylor's ideas. Since these pioneer workers first developed their startling suggestion, more and more evidence has been piling up to lend support to them.

It had been noticed 300 years ago that there was a remarkable fit between the outlines of the continents on the

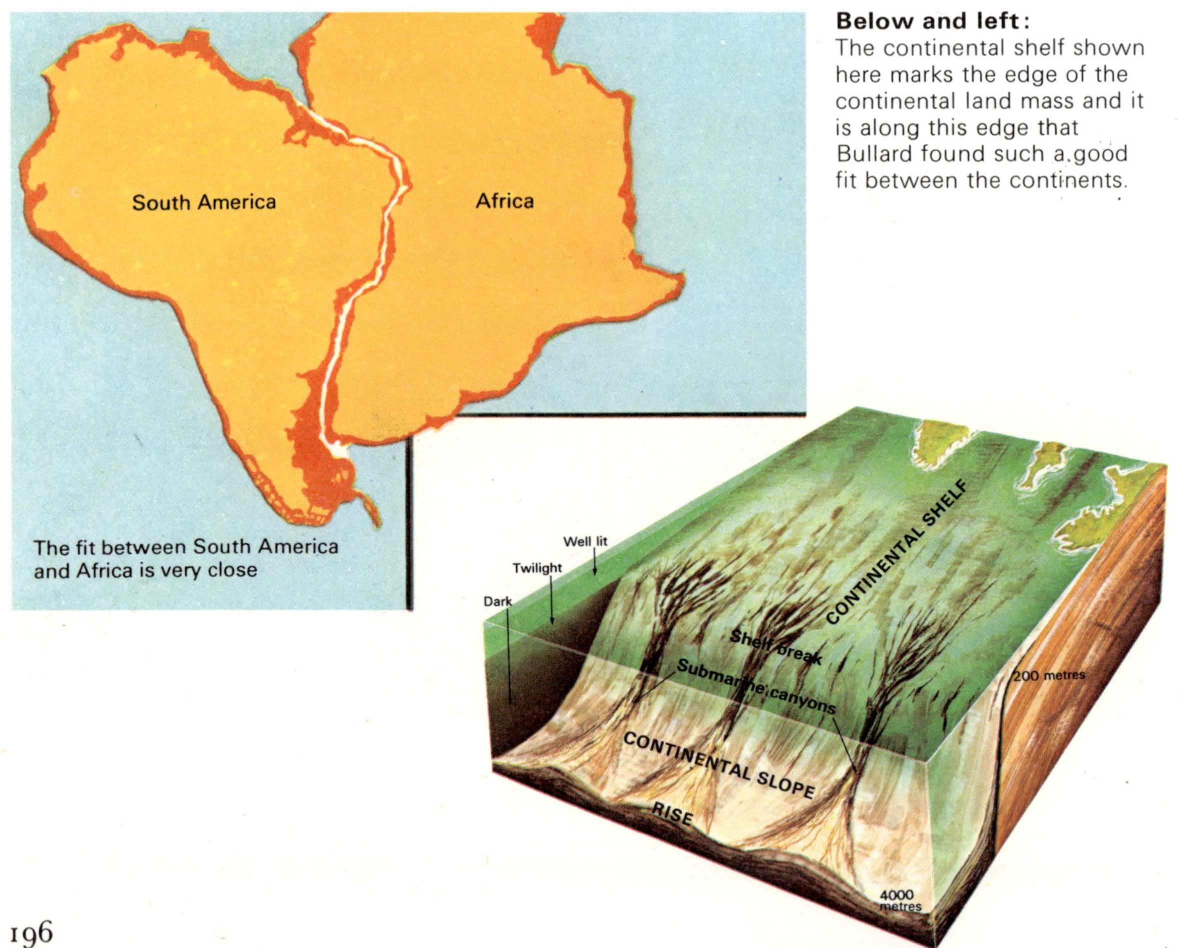

The fit between South America and Africa is very close

Below and left:
The continental shelf shown here marks the edge of the continental land mass and it is along this edge that Bullard found such a good fit between the continents.

Right:
Fossils of the same animal species found in rocks that are now separated by oceans suggest that the areas in which they are found were once much closer.

opposite shores of the Atlantic. It is particularly noticeable how well the coasts of South America and the Nigeria region seem to match up rather like two pieces of a jigsaw puzzle. An even more accurate fit was discovered by Bullard in 1965 when he used a computer to try to match up these continents along the 500 fathom (900 metre) line of their shores.

Perhaps even more surprising is the correspondence of many great formations of rock which are separated by thousands of kilometres of ocean. A particularly good example is provided by sediments deposited by ice sheets during the great Carboniferous Ice Age. Deposits of this age can be found in South Africa, India, Australia, and Assam. Now if the ice sheet had been large enough to cover these areas as they are positioned today, there would have been a world-wide fall in temperature and other evidence proves that this simply was not so. This means, then, that during the Carboniferous these areas must have formed one huge continent and that they have subsequently broken up and drifted apart.

There is yet more evidence provided by palaeontology, which although it is not conclusive in itself, taken with all else is very convincing. Between the continents of Australia and South America, it is noticeable how there are some groups of animals and plants that are obviously very similar and closely related, while others are only found in one continent or the other. It is now thought that the groups showing similarities had evolved before the break up of the continents while the very different life forms had evolved separately afterwards. It is clear, then, that the map of the world has not always been the same as it is today.

What are crustal plates?

Geology, the study of our planet Earth, is essentially a science of keen observation of all of the natural processes of the planet that go on around us all the time. Consequently, a great deal is known about the Earth, although, of course, there is still a great deal to be learned. You have already seen, for example, what makes the rain fall, how rivers form, how volcanoes work, even that whole continents are able to move, and so on. What has been needed, however, is an idea, a theory, which would knit together all that is known about the separate processes into one unified whole. This would be a theory to explain how continents drift, why earthquakes occur only in certain parts of the Earth, why there are deep ocean trenches in some areas or oceanic ridges in others.

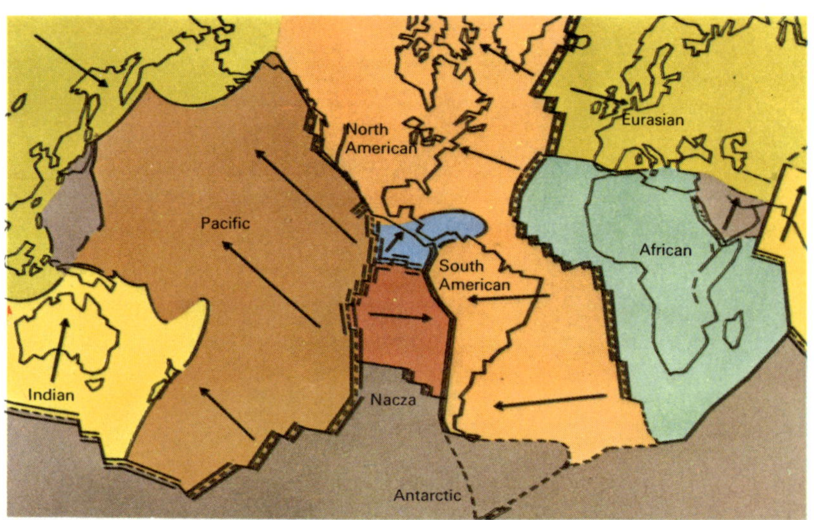

Right:
This is a simple map of the major crustal plates of the world and the directions in which they are moving at present.

Today we seem to have such a theory explaining all, or most of what has been puzzling the experts for many years. This is the all-embracing theory of *plate tectonics*. The word 'tectonics' comes from the Greek *tecton*, a builder. In other words, tectonics refers to building and structure, in this case of the Earth itself. Thus, the idea of plate tectonics suggests that the surface of the Earth is made up of a number of rigid plates comprising the thickness of the crust and the upper part of the mantle. The plates are free to move, and indeed are doing so today, floating on the plastic substance of the deeper layers of the Earth's mantle. They may well be kept in motion by the convection currents that have been mentioned before.

The important point to remember is that the plates are solid and rigid so that almost all the activities that we have

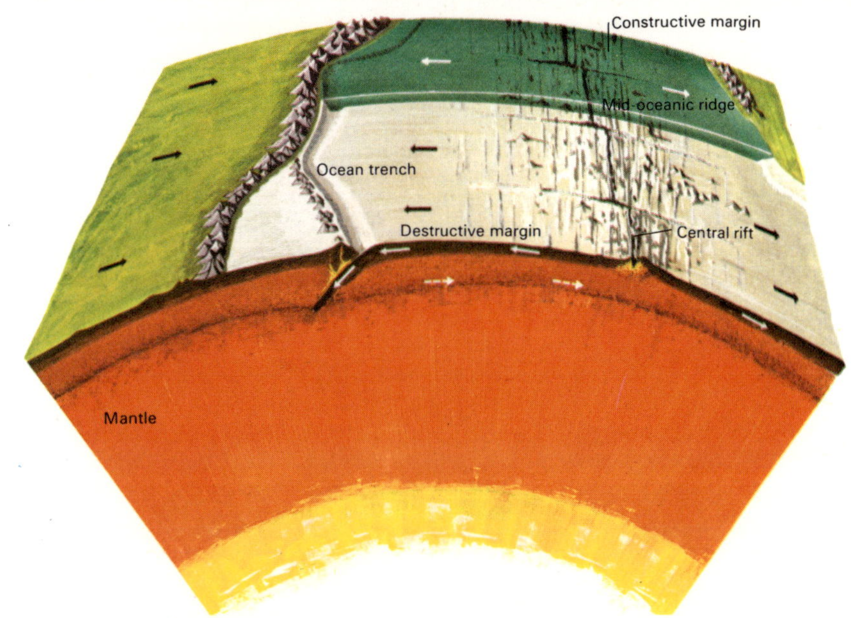

Right:
This block of the Earth's crust and mantle shows how the movements of plates, the formation of ocean ridges, mountain buildings, etc., are related.

mentioned such as volcanoes and earthquakes, are to be found associated with the edges of these plates, or the plate margins, as they are more usually called.

There are three main types of margin. The plates may just slide past each other, such as the San Andreas fault in California, and this is called a *conservative* margin. The mid-oceanic ridges, where mantle material is welling up from the depths and pushing the sea floor apart, forming new crust, are known as *constructive* margins. In other places, such as the west coast of South America, where the island arcs occur that you have already learned about, one crustal plate is slipping beneath the other at a *destructive* margin. The deep ocean trenches, such as the Marianas trench, are to be found here, where the sea floor is pulled down. Earthquake activity is often concentrated in a zone known as the Benioff zone in these areas.

You can see, then, that in one unifying idea, the Earth's activities can be viewed as a whole, rather than as a number of isolated processes.

Below:
The Earth's magnetism may be 'frozen' into cooling igneous rocks. Its direction changes periodically and these stripes record the changes progressively – older rocks moving away from the mid-ocean ridge. This is more evidence for sea-floor spreading.

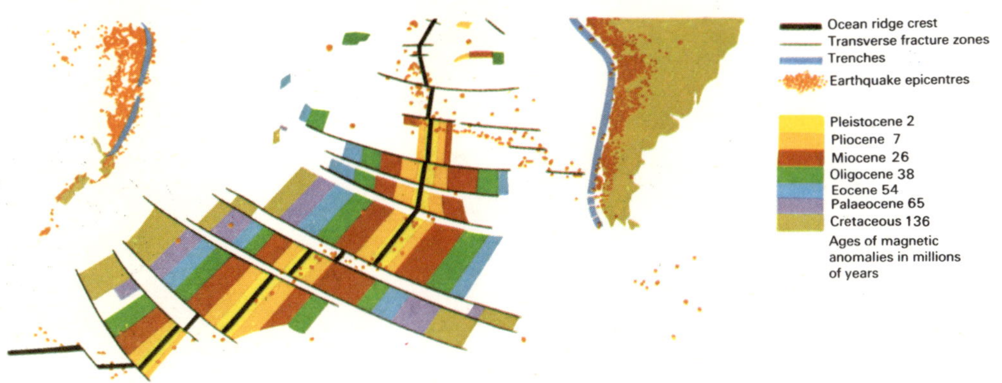

What do you know about

The Earth and Man

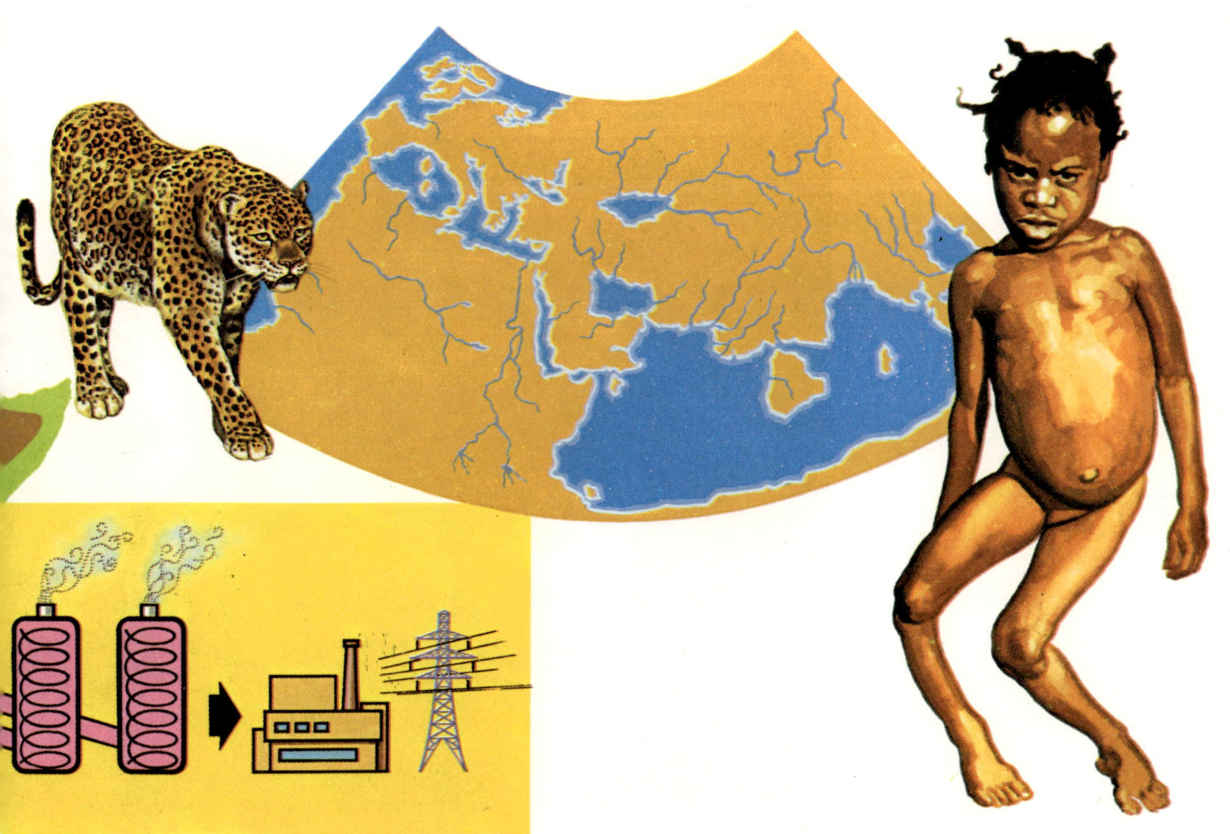

What do you need for field geology?

The answer to this question rather depends on the purpose of your geological trip. The items of equipment you take with you will vary, depending on whether you are going out into the field as a professional geologist in search of oil or minerals, or to prepare a geological map of the area, or whether you are just trying to make a good collection of rocks, minerals, and fossils, or take an interest in the land forms. There are some items of equipment, however, that no would-be geologist can do without.

When you think of a geologist going out into the countryside, we expect that the first thing that you would expect him to have is a geological hammer. Indeed, a good hammer is essential. You will need it to break off a fresh specimen of a rock for close examination (remember that a weathered surface can be misleading); you will need it, together with a selection of cold chisels, to remove perfect fossils from the rock in which they are embedded; you will need it to search in softer sands for minerals and fossils. On the other hand, it is very important to remember not to use your hammer too much; a lot of damage has already been done to important sections by careless hammering, and priceless specimens may be easily broken and lost for ever.

Right:
Some of the items of equipment that are useful for field geology. Not all of them are essential but you will certainly need a hammer, maps, compass, and a rucksack.

There are many different types and qualities of hammers on the market, from huge sledge hammers for use by an experienced geologist on hard rocks to very small, light ones. A 1½ pound or 2½ pound hammer is useful for many tasks depending on how strong you are. But remember to buy the best one that you can afford, and make sure that it is designed for the job. When some hammers are struck on to hard rock, splinters of metal fly off and you could easily lose an eye. Go to a geological supplier, ask for his expert advice and follow it!

You may also want to measure the direction in which beds of rock are dipping and at what angle. For this you will need a *clino-compass*. This is simply an accurate compass with an extra pointer. But be warned, these are expensive, and only necessary for accurate mapping. For the collector, a tough canvas bag is needed to carry specimens, newspapers to wrap them in, and a waterproof pencil to record when and where the specimens were collected. You will need a hand lens for examination of rocks, minerals, and fossils. Again, buy the best you can afford; one with ×8 or ×10 area magnification is adequate for most tasks. The last essential is a map case: whether you are just collecting or mapping you will need maps and they must be protected from wear and water. Map cases are readily available, but you may find it better to make one to your own design.

How do you put a mountain on a piece of paper?

Compared to the size of the Earth, man is minute. Compared to a mole hill in a meadow, man is much bigger. We can look down on it, and see it clearly. If you were the size of an ant, however, and you had not climbed that particular mole hill before, it might not be quite so easy. In man's case, we would find a mountain that we had never seen before as difficult to explore as an ant would a mole hill. We need something to help us find our way over the mountain without taking the risk of getting lost or falling over a dangerous precipice.

Right:
This map shows the 'inhabited' world known to the ancient Greeks. The Greek geographers made a distinction between their limited knowledge of the inhabited world – the home of man – and the planet Earth.

Some animals are able to find their way by smell. When a badger goes exploring, for example, it leaves a scent trail behind it. If the badger wants to re-use the path it simply follows its nose. Some birds are even more clever at finding their way from one place to another. Swallows for example, travel hundreds of kilometres from Africa to Britain and Europe each spring to lay their eggs and rear their young, and then fly back again each autumn. We do not know exactly how birds migrate huge distances without getting lost, but it is likely that they navigate using the Sun and the stars to guide them. It has even been suggested that they are able to detect the Earth's magnetic field to help them find their way.

Man is not so clever. If you did not know your way from one place to another, you would look at a *map* to help you. A map is simply a way of showing the three-dimensional countryside on a small scale in two dimensions.

What is a typical map? In Britain, for example, a typical map is an Ordnance Survey map. If you look closely at one, you will notice that the map is drawn to a scale. It could be

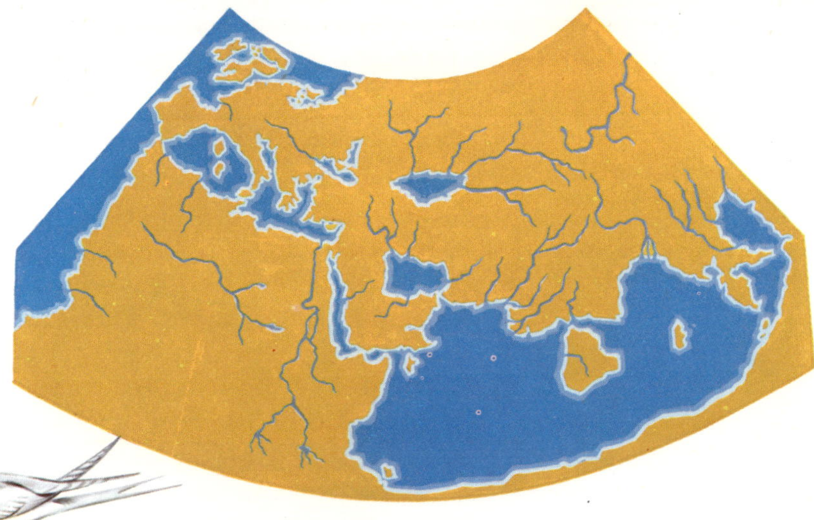

Right:
In the second century AD, the astronomer Ptolemy made a map of the world which included all the geographical information then available.

Above:
Each year the Arctic tern flies more than 40,000 kilometres as it journeys between its northern nesting areas and the south. It manages to find its way without the aid of a map.

1:50,000, for example. This means that something which measures one kilometre from end to end in the area shown on the map would measure two centimetres on the map sheet.

What is shown on a map? Of course, this depends on the purpose of the map. But look at your Ordnance Survey map again. You will see roads represented by red, orange, or yellow lines depending on how big they are. You will see railway lines marked by black lines, either dashed or uninterrupted. Rivers, lakes and the sea are shown in blue. Woods are shown by areas of green. Then, of course, there are a number of symbols to indicate such things as churches or hotels, quarries or telephone boxes and much more besides. Perhaps one of the most interesting things you will notice on the map are the faint orange lines that seem to form irregular loops. They are called *contour lines*. Each contour line marks the particular height of the land above sea-level. They are usually marked at intervals of 50 feet. It is these contour lines that mark the position, extent, and height of a mountain on the piece of paper.

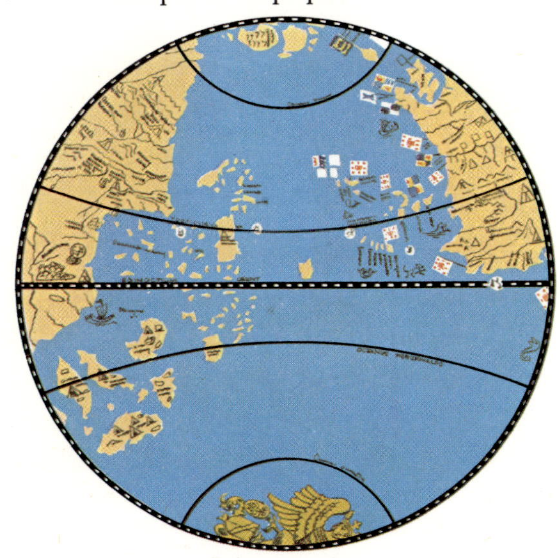

Right:
This globe was completed by Behaim, a German geographer, in 1492, the year that Columbus discovered America.

What is a geological map?

If you wanted to show all the different kinds of rocks in an area, their age and order of deposition, and the structures into which they had been deformed by Earth movements, you would do so by means of a geological map. We have said in the previous question that a map is in two dimensions on a flat sheet of paper. This is usually the case, but there are *relief* maps which are attempts to show the actual appearance of an area of country on a small scale. In other words, relief maps could be thought of as a model of a particular area of countryside.

We expect you have already seen a simple geological map, especially if you have been to your local museum which will always have geological information about your locality. To the untrained eye, a geological map looks like a sheet of attractively coloured paper, but to the geologist or even to the person with just a little, basic geological knowledge, this map can be a source of an enormous amount of information about the area it covers. If you have a rough idea of how a geological map is prepared, then this will give you a clue to the kind of information that you can learn from such a map.

The geologist will usually begin with a *topographical* map (giving detailed information about the surface outline of the area, that is, showing by means of lines of equal height, the *contours*, of all the hills and valleys). This map will usually be

Right:
A simplified geologic map of Wales where much pioneer work in geology was carried out by such men as Charles Lapworth and Adam Sedgewick.

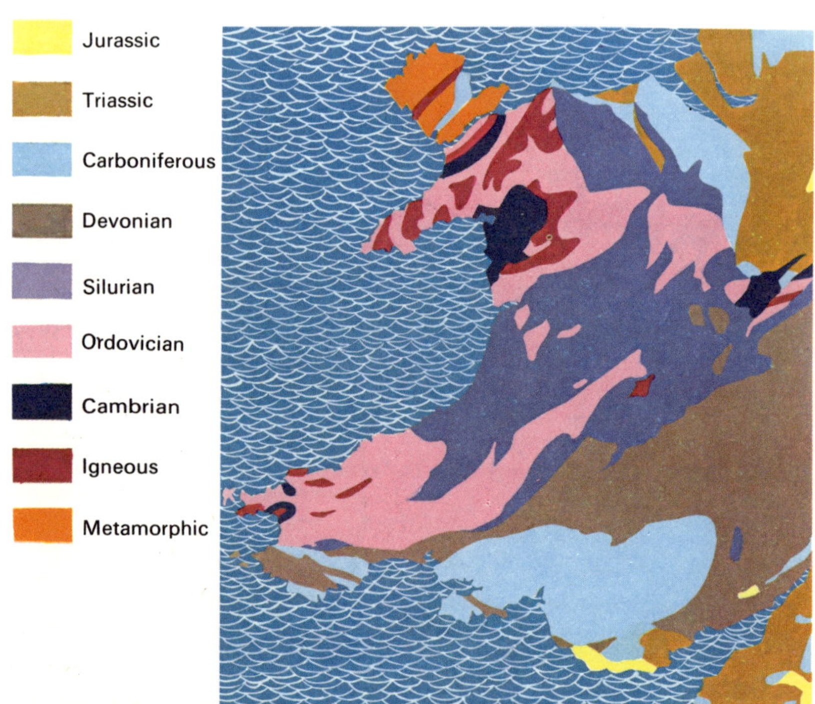

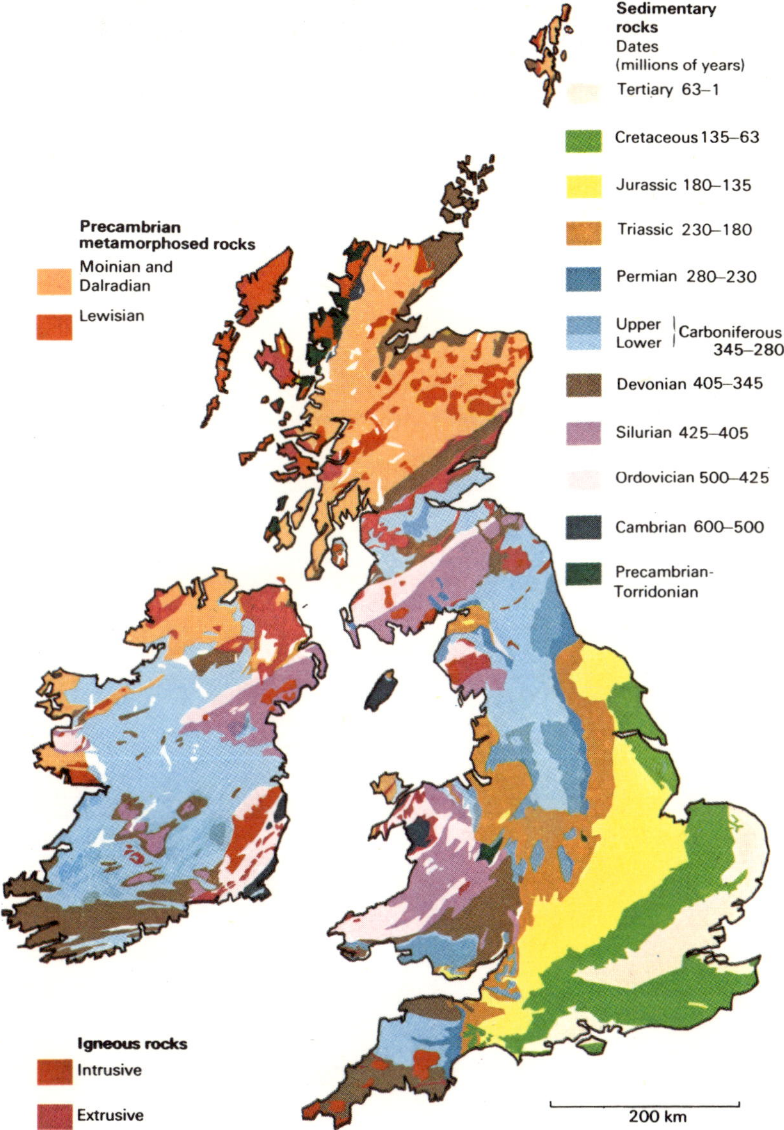

Right:
A simplified geologic map of the whole British Isles. You can see that there is a great variety of rocks leading to the interesting and varied scenery to be seen in a very small area.

to a scale of 6 inches to 1 mile. Then comes the hard part for the field mapper: he or she must walk over every part of the area and record on the six inch sheet every lump of solid rock that is showing at the surface. The result is an *outcrop* map. If the area has already been mapped, then the geologist will have an idea of the order in which rocks were laid down and/ or igneous rocks emplaced, and even their accurate ages. If the field mapper is in the area for the first time, this will have to be worked out by using the right way up structures that we have already mentioned. You will remember that when sediments are deposited, structures are formed (such as current bedding) which will indicate the way up of the original sediment.

How do geologists help us?

All the sciences, mathematics, physics, chemistry, and biology, are playing an increasingly important role in our way of life, as the years pass and we become more dependent upon the products of science and technology. Geology, too, has a vital part to play in living today, apart from its value in furthering our understanding of the way in which our Earth began, developed, and is in constant motion to this day. For example, few of us could imagine living now without the vast network of roads and railways that span the land surface. But these roads and railways require bridges and tunnels which cannot be built in some areas or they may collapse.

Right:
It is very important that the rocks that support huge structures like this dam must be strong enough to withstand all conditions.

By careful surveying, the geologist is able to tell what kinds of rocks there are in an area, and can suggest to the engineer the best site for their construction. Before geologists were consulted, huge constructions were often erected on sites quite unsuitable for the purpose with disastrous results. A famous instance is the San Francisquito Dam disaster in America. A quick survey was carried out on the proposed site during the dry summer months and it was decided that the narrowest part of the river channel was ideal. Unfortunately, the rocks on one side of the canyon were clays, with schists on the other. When the water filled up behind the dam the clays became soft and flowed away so that the schists broke away and the whole dam collapsed.

Right:
By careful geological survey disasters like this one can be avoided.

Geologists can also help to avoid the terrible loss of life which occurs as a result of earthquakes. So far they have not been able to perfect methods of actually preventing earthquakes, but at least with the methods now available (remember 'What makes the ground tremble?') people can be warned of a coming disturbance in time to escape.

Perhaps today the geologists' skill in finding raw materials is even more important. Industrialized society depends on the raw materials that the Earth provides – nowhere else can provide our needs. Our society with its demands for more energy, more water, more iron, more aluminium, more chemicals from oil, and so on, now asks the geologist to find more sources for all these materials and many others. Obviously it is becoming more difficult and more costly. In time, of course, unless our demands are reduced, not even the cleverest geologist will be able to find new supplies. They will have all been used up.

Right:
As geologists find it more difficult to find new supplies of fresh water, distillation plants which produce fresh water from sea water will prove to be more and more valuable, particularly in areas such as the Middle East.

How do we look for the raw materials we need?

You have already seen that the geologist has an important part to play in the modern world; one of his most vital functions is to find deposits of raw materials for our industries. The deposits should preferably be in a place which is easy to get at so that materials can be transported cheaply. They should also be in concentrations that are rich enough to allow the mining or oil company to remove them at a profit. In fact, deposits of this kind are very few and far between because they have all been used up. The geologist must work harder and harder using better and better techniques. But what are the methods of the modern exploration geologist?

Right:
Gravity survey can show, for example, that a mountain has a 'root' of lighter material than the denser mantle rock.

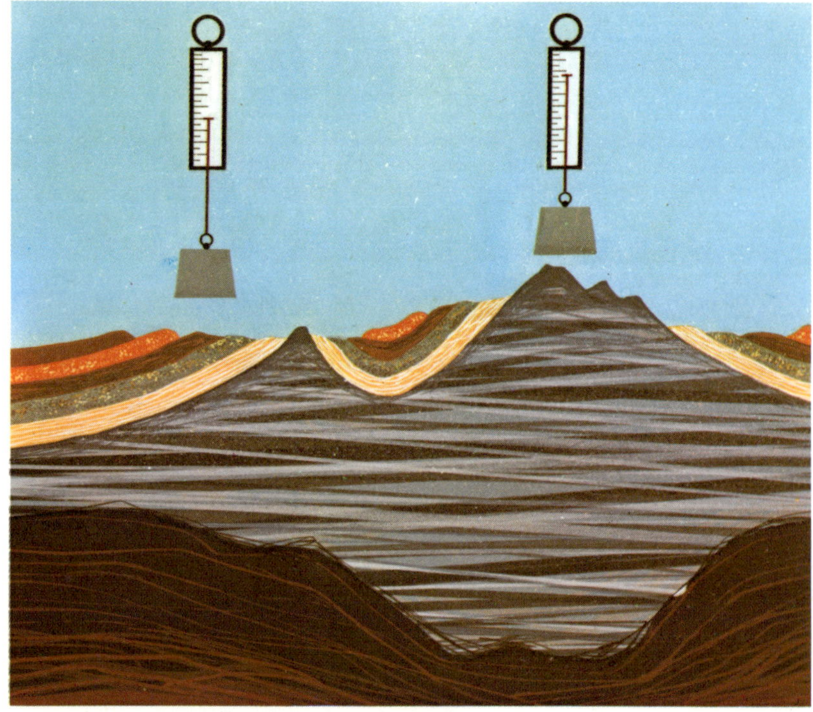

Below:
The old-fashioned method of panning for valuable heavy metals is now passing into disuse.

The geologist's method depends on what he is looking for and the type of countryside he is exploring. It cannot be stressed too strongly, however, that with all the advances that have been made, there is still no real substitute for the geologist's experience and ability to prepare a really accurate geological map. (You have already seen how a map can be prepared.) If the geologist is looking for minerals or ores such as iron, copper, or tin, a chemical analysis of the soils and streams in an area can provide very valuable

information. For example, if he takes samples of the water from selected points up a stream, he may find that there are traces of copper in the water in the lower reaches and that much further up-stream there are not. He knows, then that the source of the copper must be between those two places. It is possible to track the origin down quite closely in this way.

You have already seen that when an earthquake occurs, the waves produced travel through different rocks at different speeds. This property can be used on a much smaller scale. If a shock is produced in the ground with a hammer or explosives, measuring instruments can record the waves at selected points and a good idea can be gained of the geology below the surface. It is also worth noting that the pull of the Earth's field of gravity is not constant over the whole of the Earth. It varies with the distance from the centre of the Earth and with the geology of the area which is measured. Denser rocks exert a stronger pull than lighter ones, and sensitive instruments can be used to measure these tiny differences.

A direct look at the rocks below the surface can be made by drilling, and even when oil wells have proved to be dry they are not completely wasted because they can reveal a great deal about the geology. An instrument like a giant corkscrew called an *auger* can be used by the worker in the field for very shallow investigations.

Above:
Even with all the new survey methods, the experience of the geologist's eye is still important.

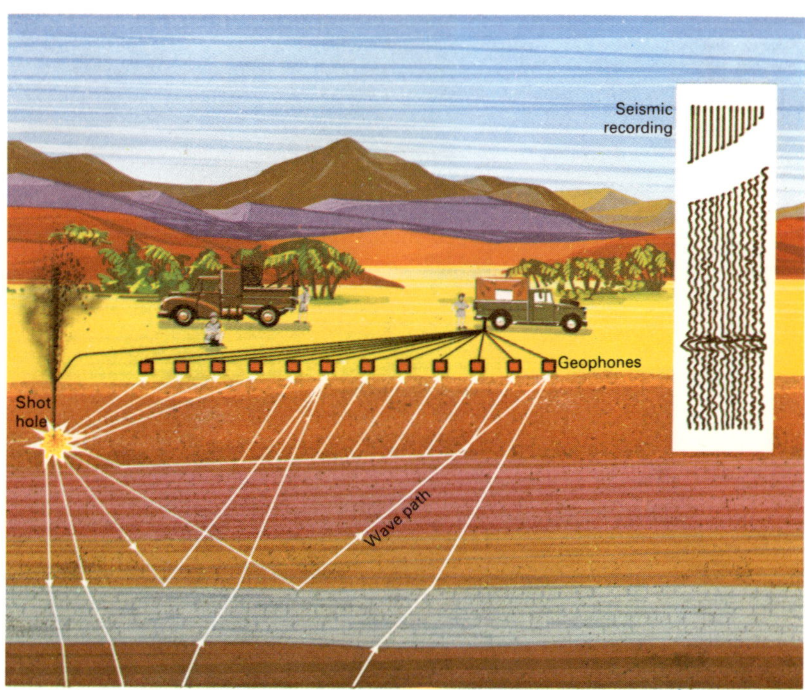

Right:
This type of survey takes advantage of the ways in which shock waves travel through the various rock layers.

What are ores, and where might you expect to find them?

Above:
This typical poor quality iron ore can be found in western Australia.

You must realize that the Earth provides us with all the raw materials swallowed up by our industries. Many of these materials are minerals. The meaning of the word 'mineral' has already been explained, but sometimes it is used to denote any substance obtained by mining. It is clear that the words mine and mineral have similar origins. You might expect that minerals would be spread throughout the Earth, and indeed, this does happen. Occasionally, however, (and fortunately for our present needs) minerals may accumulate in sufficient quantities for man to be able to remove them comparatively easily. When this situation arises, the deposits are usually known as *ores* or ore bodies. The minerals associated with the ores that have no real economic value are usually referred to as *gangue* minerals. Today, because of the greater demand for minerals, the consequent scarcity of them, and improved methods of mining and refining, material that was once thrown on the spoil heaps can now provide valuable sources.

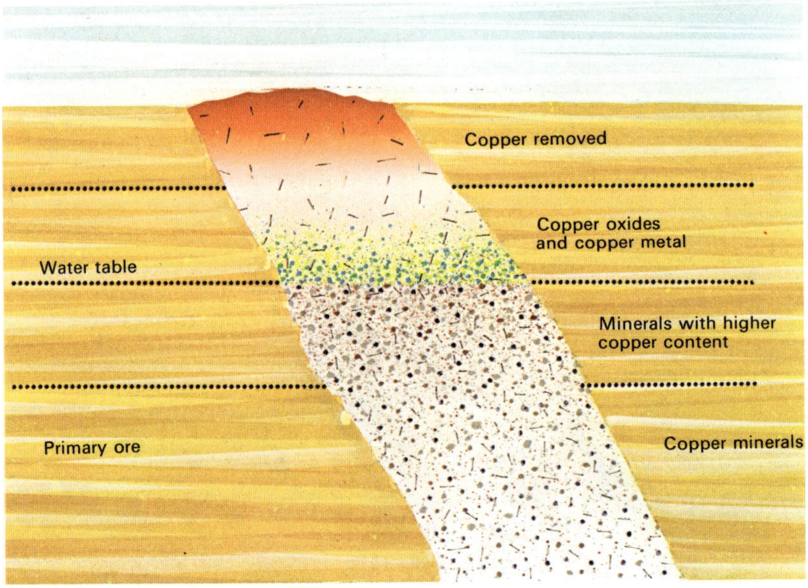

Right:
The concentration of a metal such as copper can be changed by the movements of water and by weathering.

Ores can occur in a variety of ways. With ores of iron, for example, the same ore can arise as a result of different methods of concentration. Many ores other than iron are formed in association with magmatic processes (you have seen the word magma before – look at the questions on igneous rocks if you cannot remember it). If you remember the way in which a magma cools and forms crystals, it will come as no surprise to you to learn that metals such as chromium and nickel result from the settling into bands of the relevant crystals as the magma cools. There are often watery solutions charged with minerals coming from magmas and these also provide supplies of metals such as

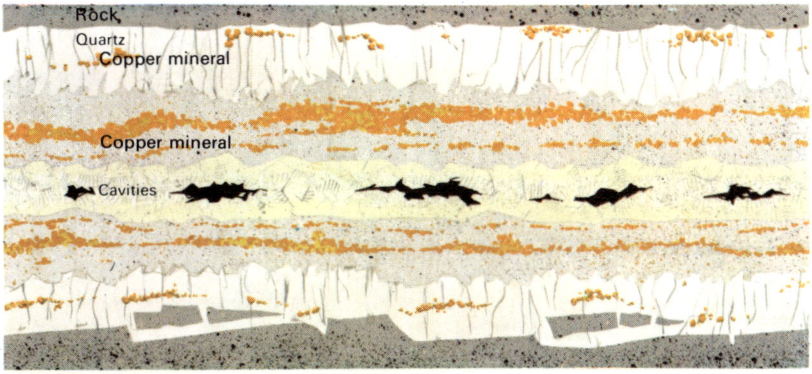

Right:
A banded copper ore vein.

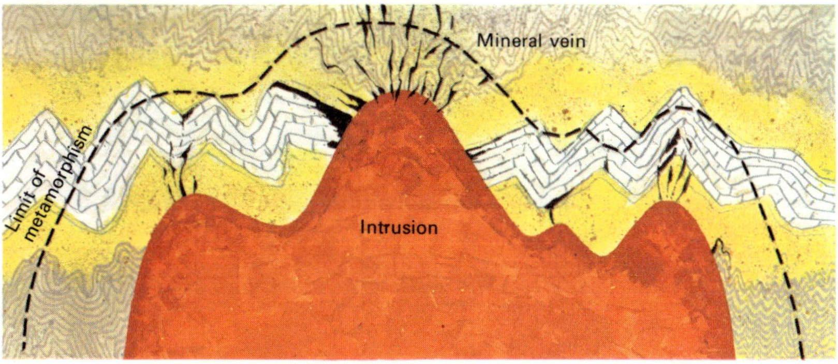

Right:
Deposits of ore may be associated with large igneous intrusions.

mercury or copper. Both these minerals are in short supply, and in fact, deposits of mercury are almost completely confined to areas in Spain. Not all ores have been formed by igneous activity, however. Important deposits of iron in England have resulted from concentration by sedimentary process. Deposits known as *placers* are typical sedimentary ores. Of course, oil and coal must be considered as economic minerals and these have obviously resulted from sedimentation.

The sea may one day provide much of the world's minerals. It has been known for some time that there is great richness of gold in the sea but as yet it has not been worthwhile nor even possible to exploit this mineral wealth. Even uncommon metals like vanadium, which is used for the nosecones of spacecraft, is concentrated in the blood of an unexciting sea animal – the sea cucumber.

What is ecology?

You have seen the way in which the first primitive forms of life first arrived on our planet, how they multiplied and became more complex. You have learned something about the main animal and plant groups, and in particular, about those fossils that are of special importance to the geologist. You have even looked at the origins and the evolution of man himself. Each animal or plant has been considered separately, however, and only in terms of its biology and evolutionary history. You know, of course, that no living organism lives a totally independent and isolated existence. Even man depends on other animals and plants for food, and clothing; even the fossil fuels come from living things long dead.

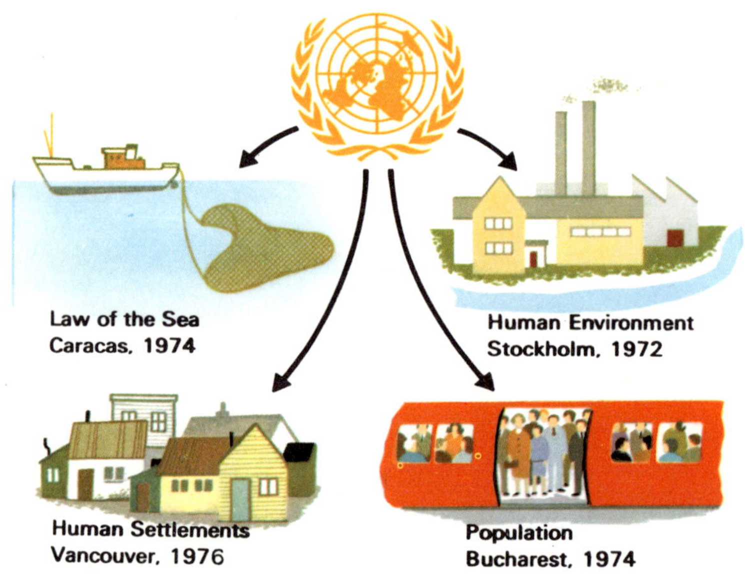

Right:
People and governments are becoming more aware of the importance of protecting our natural environment, as these recent international conferences indicate.

The study of plants and animals in relation to their environment (and this includes other animals and plants) is known as ecology. The word comes from the Greek *oikos* meaning the home and *logos* meaning discourse or speech. You probably do or have done nature study. Ecology really just applies scientific methods to this basic study. Instead of considering just animal life as does the zoologist, or plant life as does the botanist, or where animals and plants live on the Earth as does the biogeographer, the ecologist must have a working knowledge of all these and many more sciences in order to come to an understanding of the nature of all living things. The ecologist must look at the way all animals and plants depend upon each other for their very existence in a complex natural web.

Right:
Part of a hedge bottom showing just a few of the plants and animals that live in this very rich habitat.

The ecologist does not confine himself or herself to other life forms. Man is an animal and must be looked on as such. Perhaps it is worth remembering that while all animals try to change their environment to suit their own needs, man is able to exert a much greater influence in his attempts to improve his lot. Man has been able to achieve this because of his intelligence and his ability to make and use tools – once just simple stone hammers and axes but now computers and guided missiles. A song thrush may use a particular stone time after time to break the shell of a snail to remove the soft body inside. Thus, it is making use of a tool. A badger may build its sett in a bank and line it with dry leaves to keep itself and its young warm. Thus it is making a home. Man does all these things but on a larger scale, more quickly, and more permanently.

You can see then that the ecologist's knowledge of living creatures could become more and more vital to man as he needs to grow more food and find room for more and more people.

Right:
A small pond can support many different plants and animals dependent on one another for survival.

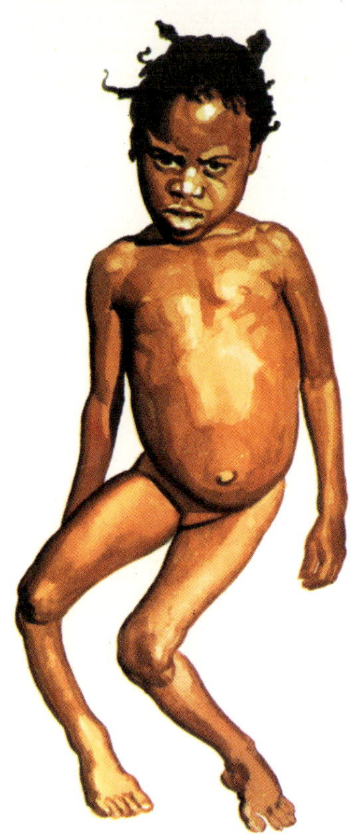

How do animals and plants depend upon each other?

The sun is the driving force in all of the Earth's processes. It is the sun's rays that keep our planet warm enough for us to be able to survive, it is the weather's motor, and provides energy for the most important process of all so far as plants and animals are concerned. The sun provides plants with the energy for *photosynthesis*. It is by photosynthesis that plants use the sun's energy to convert carbon dioxide taken in from the air, and water from the soil into the sugars and starches that make up their stems, leaves, and roots. It is a little more complicated than this, and plants need other substances as well, but this is the important process. Photosynthesis only occurs in plants that contain the complex green pigment *chlorophyll*, in other words in green plants.

Other types of plants depend on other means of manufacturing these materials. It is interesting to note that green plants can store energy so that photosynthesis can even occur at night. You have seen, too, how the first plants evolved in the sea. This is because water is needed to provide the hydrogen to form the sugars and starches.

Above:
We all depend on our natural environment. This child is suffering from a disease called rickets, a softening of the bones due to lack of a certain vitamin and sunlight.

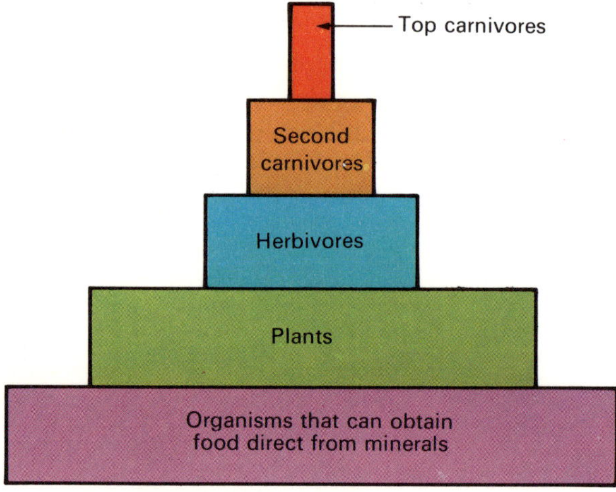

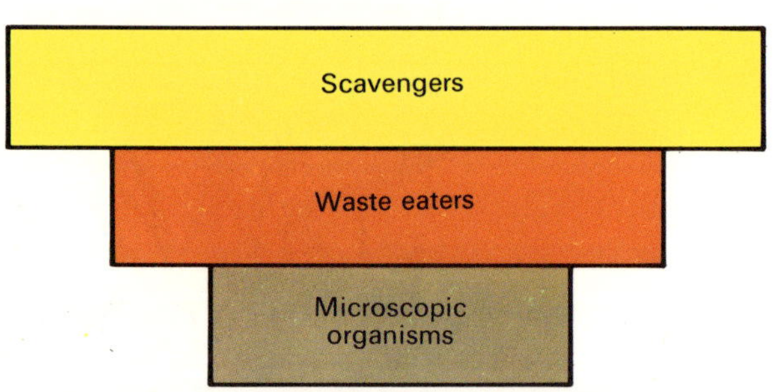

Right:
Carnivores, animals that eat other creatures, depend upon animals that eat plants, herbivores. Herbivores in turn, depend on creatures in the soil that release food for them. There must be fewer carnivores than herbivores, and so on.

Left:
A cat might eat a blackbird which might feed on ladybirds which might depend on greenfly which lives on a leaf. This is called a food chain.

We now have the basic food supply for the rest of the planet's life forms. You know that cows, for example, eat grass. Without grass or other plant material available, cows could not survive. We then make use of the cow's ability to digest grass and convert it into meat and milk which we eat and drink. Thus, the cow is a herbivore, that is, an animal that feeds on plants. When we eat meat, we are behaving as carnivores or meat eaters. This demonstrates a simple chain of events, in fact, it is known as a *food chain*. The grass grows and is eaten by the cow which is eaten by man.

There are many other such chains. A fox may feed on rabbits, which may be eating clover. The clover may be dependent not only on the sun, water, and the air, but also on bees for pollination. These chains can be disturbed quite easily sometimes with far reaching effects. For example, should bees become reduced in number for any reason, such as a farmer using insecticides, in time the amount of clover would fall back and the rabbits might be forced to feed on the farmer's lettuces. Obviously, the farmer will not be happy about this and may set about reducing the rabbit population. If he was too enthusiastic and all the rabbits in an area were to be wiped out, the fox would have to look elsewhere for its meals – it might even be the same farmer's chickens.

Try to imagine multiplying these simple chains thousands of times to take the multitudes of different animals and plants into account. Now you begin to have some idea of the complicated nature of life on our planet, how animals and plants depend upon their surroundings and each other, and how easily the balance can be upset.

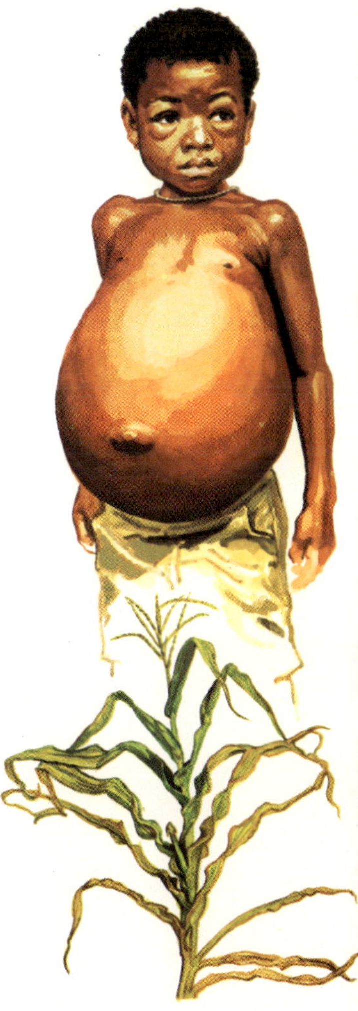

Above:
This child is suffering from a lack of protein. Even plants require the correct nourishment or they will become diseased.

How can you help to look after our Earth?

The first thing to remember is that human beings are animals. Admittedly, humans are very intelligent and adaptable animals, but animals just the same. Our species has evolved on the Earth as part of the Earth's web of living things. We were not born somewhere else in the Universe and then landed on Earth later. This means that we are adapted to living in the conditions which the Earth provides for us. After all, any animal that is not fully adapted to its surroundings or which cannot compete with its rivals in the struggle for food and shelter eventually ceases to exist. We depend upon the energy provided by the sun to keep us warm, we need water, and we need food in the form of vegetable materials or other animals. If the time came when we were the only living species on the planet our days would be numbered.

You can see, then, that in order to protect ourselves we need to look after the Earth and all its creatures. To do this the first step is to try to understand a little more about the way in which living things behave, and to try to appreciate the beauties of the world. This requires us to go out into the countryside and watch and listen. Unfortunately, by going into the areas that we are seeking to protect it is all too easy to cause considerable disturbance. When you go into the country, remember that it is living and growing, and follow the Country Code:

Below:
If areas of countryside are burned repeatedly, the soil can be eroded away leaving desert conditions.

Above right:
Carelessness in the country-side can lead to forest fires killing thousands of plants and other living things.

1. Guard against all risks of fire – do not, for example, throw down lighted matches or drop any glass which may act as a burning lens, and try to see that other people are just as careful.

Left and below:
Nature reserves like the three shown here can help to protect areas of outstanding beauty or plants and animals that are in danger of extinction.

2. Fasten all gates – animals straying into the wrong field may eat grass that is too lush and become ill and even die.

3. Keep dogs under proper control – a dog may easily frighten domestic or wild animals, and this is even more serious if it is during the breeding season.

4. Keep to the paths across farm land.

Below:
Marshy lands at river estuaries are particularly rich in wild life. These areas are decreasing because the land is suitable for reclamation by man.

5. Do not damage walls, gates, hedges, or fences.

6. Take all your litter home. Animals can be cut by broken bottles or cans, or choke on plastic bags.

7. We need water – look after all possible sources.

8. Protect all wild life, plants, and trees – there are laws to protect many species of birds, plants and animals that are in danger of dying out.

9. Go carefully on country roads – if you are in a car, particularly at night, it is all too simple to run over a hedgehog or a fox.

10. Respect the life of the countryside.

These are just a few of the points to remember, but before you do anything, just stop to think of the consequences *before* you do it.

Which parts of the Earth are in most need of protection?

There are certain areas around the globe that are particularly vulnerable to disturbance whether it be by man or such factors as climate. Some areas, such as deserts or the ice caps of the Arctic and Antarctic have definite characteristics of their own, but what of the areas between? Take, for instance, the boundary between the oceans and the dry land. This is where you would expect to find the estuaries of rivers and coastal wetlands generally. In the past, these areas have been considered to be of little value. It was thought that by reclaiming them from the sea, the land could be put to better use whether as farming land or land on which to build a new airport.

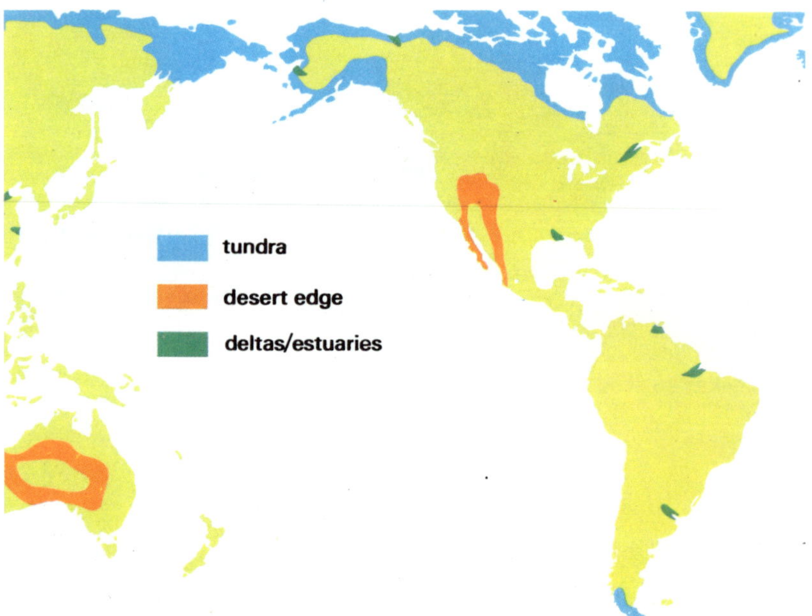

Right:
The maps on these two pages indicate the areas of the world that are most vulnerable to any changes of the environment. They are areas between the larger, and more stable geographic zones and many of the plants and animals that live there could not survive anywhere else.

In reality, however, these regions are not the waste lands that they at first seem to be. We depend upon the fishing industry to supply us with food and fish products for fertilizers and so on. Many sea fishes spend much of their development period in the estuaries, and if estuaries were to be destroyed, it would not be long before the stocks of fish became exhausted. Even animals like prawns depend upon the very rich supplies of food in estuaries to survive.

The mudflats surrounding river mouths provide feeding grounds for an important section of the bird community, the waders. Waders include birds such as the redshank and the dunlin, the oystercatcher and the godwit. You have seen

Right:
The southward spread of the Sahara desert means the loss of food supply to people that live close to this area with obvious consequences.

in the previous question what can happen to a food chain if it is interrupted in some way. It is important to remember that the destruction of any species of animal or plant by any but purely natural causes does not only remove something beautiful from the Earth but may even affect our own food supplies.

On a somewhat more local scale, it is important that our farmland should be used sensibly, and protected to supply our needs for generations to come. It is easy to see the effects of carelessness. With the coming of the tractor and then even larger and more complex farm machinery such as the giant combine harvesters, it seemed that the bigger the fields, the easier and cheaper it would be to manoeuvre the machinery. As a consequence, many kilometres of hedges were uprooted, to remove obstacles to the progress of the combines. This meant that a great deal of important habitat was lost to wild life. In addition, there was a more direct result. It had not been realized that these hedges acted as windbreaks, and their removal led to widespread soil erosion. The lesson to be learned once again is to think of the consequences before acting.

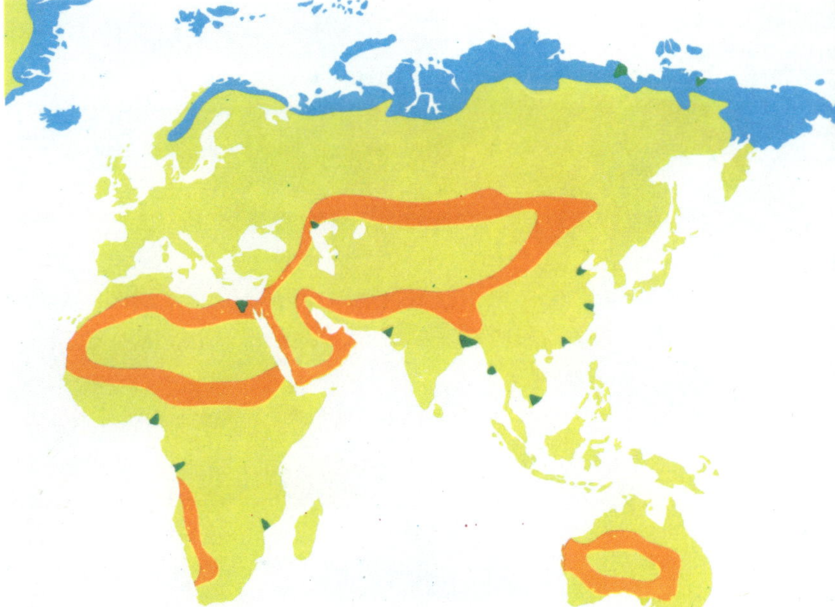

Right:
Notice that those areas that are particularly vulnerable to changes in the environment are generally relatively small. They include deltas and river estuaries.

Which animals and plants are in the most danger?

Below:
Here are a few of the many wild flowers that are rare or threatened with extinction. There is little co-operation between countries to conserve species. Laws protecting wild flowers are often difficult to enforce.

In this question, the word 'danger' does not refer to the danger to which individuals of any species may be subjected. Danger in this case means the danger to which a whole species is liable. In other words, species may be in danger of becoming extinct, and disappearing for ever from the face of the planet. Of course, we can tell from the fossil record that hundreds of species of animals and plants have come and gone throughout the span of life on Earth. In the past, however, those species that have died out have done so as a result of natural causes. The animals and plants that disappeared were those that were unable to adapt fast enough to changing conditions or were unable to compete with their rivals for the available food supplies. There are instances, however, such as that of the dinosaurs, in which it is not at all clear why they became extinct. Many ideas have been suggested, but none of them is wholly satisfactory.

For today's wild life, there is another, more terrible threat to their existence – man. Man is endangering the lives of many species of animals and plants which are unable to adapt to the pressures of man's increasing numbers and his changes to the natural environment. On the other hand, there are those species that have certainly benefited from man's activities. In Britain, for example, birds such as the black headed gull have increased in numbers quite dramatically in the last fifty years, adapting well to a semi-urban environment and feeding on man's waste. Even the fox has been able to survive by adapting to become an urban scavenger in many large cities. Unfortunately, the lack of suitable foods has led to the foxes becoming deformed so that individuals probably spend a great deal of their lives in considerable pain.

But man is threatening wild life in many ways. Firstly by his spread further and further into the rural areas, man is reducing the areas in which wild life can live. This also includes the effects of widespread farming where forests and heaths come under the plough and many kilometres of

hedges are removed. Secondly, man has hunted many animals to extinction. A good example of this is the destruction of the great auk during the last century or perhaps even more famous the death of the dodo. More recently the fate of many species of the giant sea mammals, whales, has been in the news, as their numbers fall to the more efficient whaler's harpoon gun with its explosive harpoon. Pollution is also affecting the lives of many animals. Millions of sea birds die very unpleasant deaths as a result of their feathers becoming covered in sticky, black oil waste.

It is sometimes argued that if an animal or plant cannot survive man's ravages, then this is just the course of evolution. Unfortunately, many beautiful and valuable species, on which man himself depends, may never be seen again.

Above:
Many wild animals have been hunted almost to extinction and even put to horrible deaths to satisfy the demands of fashion.

Left:
This bird, the great auk, was the northern hemisphere's counterpart of the penguin. Egg collecting and hunting drove it to extinction in 1844.

Who helps to look after the Earth?

You have seen how much we depend upon our planet and its other inhabitants in so many different ways, and although this seems obvious enough, most of us take our environment very much for granted. How many of us stop to wonder what the products of industrial living have cost in terms of loss of countryside or polluted oceans or dead animals. The need to be concerned about the delicate balance of nature cannot be stressed too strongly, although some people feel that because man is as much part of the natural world as any other creature then his widespread interference with the systems of life is acceptable. There are others who lack concern completely.

Right:
Seabirds like the guillemot are in most danger of oiling because of the way in which they feed. They dive under the sea and then surface some metres away, perhaps into an oil slick leaked from a tanker or an oil rig. Volunteers save the lives of many birds by cleaning them with detergent.

It may not yet be time to despair, however. All over the world there are men and women, scientists and non-scientists who have taken it upon themselves to act in such a way as to protect some part of our environment, often at considerable personal cost. In Britain, for example, there are many organizations that are concerned with various aspects of the living world. On a national scale there are bodies such as the British Trust for Conservation Volunteers to which people are able to offer their services for many kinds of conservation work. No experience is required, and volunteers learn a variety of skills from pond clearance to tree planting. The gatherings are usually arranged for week-ends, and during the holiday periods residential trips are arranged at low cost. On a more local scale in Britain, there are the county naturalists' trusts such as the Berkshire, Buckinghamshire, and Oxfordshire Naturalist Trust or BBONT for short. The main aim of these groups is to obtain, whether by gift or by buying with their own funds, areas of particular natural interest which can then be protected as reserves. It is hoped that in this way many of the animal and plant species that might not otherwise survive will be given a better than even chance. This means that these groups must try to have as many of the different types of habitat as possible and must have funds enough to employ people to manage them. There are of course the better known groups that are concerned with particular types of species. The Royal Society for the Protection of Birds is one of these.

Right:
Animals such as the elephant that were once mercilessly hunted just for their ivory are now afforded some protection in national parks and game reserves but these areas have to be looked after very carefully.

The Friends of the Earth are concerned with all things environmental. They are a limited company rather than a charity which means that any actions that they take can be completely independent. Recently they have campaigned against such things as the use of no deposit bottles for beer and fizzy drinks and have been quite successful in their efforts to save the whale from extinction.

These are just a few of the people and organizations that are concerned with looking after our planet, but there are still many ways in which we all can help.

What dangers does the Earth hold for its peoples?

We are intelligent animals that have evolved on the Earth. We have adapted to suit its requirements and we have tried to change our own environment to make our lives more comfortable with varying success. The Earth is our provider. At times, however, it is also our enemy.

The dangers which the Earth holds for its peoples depend to a large extent upon which parts of the globe we are talking about – the areas where there are extremes of any kind are the most dangerous! In talking about some of the processes of the Earth, we have also mentioned the terrible forces locked up and the havoc which they can wreak. We have, for example, already mentioned earthquakes. Scientists watching the earthquakes in Chile in South America described the earth beneath their feet as 'slow and rolling like that of the sea during a heavy swell'. Can you imagine how terrifying that might be? In very thickly populated areas such as Tokyo in Japan, the people there live their whole lives under the threat of an earthquake which could send the whole city toppling with awful loss of life.

Although earthquakes are very dramatic, they are not the only natural disasters which threaten life on Earth – the weather can also play its part. In tropical areas close to the oceans, whirling winds can tear down buildings, rip up trees,

Right:
People living in this area might suffer with drought for most of the year, and when it finally does rain, flooding might take place (*see* facing page).

and pick up motor cars as though they are toys. They are caused when the air has been heated more than normal, sending up twisting streams of air. Around America, they are usually known as hurricanes and around Australia and Japan they are referred to as typhoons. They usually affect quite small areas, usually no more than 400 kilometres across, so that the air pressure gradient is very steep and in the eye or middle it is quite calm, but around this, winds may reach speeds of 350 kilometres per hour. There is also the tornado which usually occurs on land and may begin as an off-shoot of a hurricane. This is a very narrow column of spinning air, perhaps no more than a kilometre across, but causing dreadful damage to all in its path.

In the temperate regions of the Earth such as Britain and Europe, there is generally a good steady supply of rain and temperatures that are neither very low nor very high. In some hot dry areas, such as India and Pakistan or Ethiopia, they are very dependent on the rains for their water for drinking and irrigation of their land to grow food. Sometimes the rains do not come, the people cannot grow food, and thousands die of starvation, the very old and the very young suffering the most. When the rains do come, they may be so intense that flooding occurs, making the situation even more horrifying.

These are just a few of the ways in which the Earth can endanger us, but there are many more.

Above:
Here are three more ways in which the Earth can endanger us:
(1) The tremendous winds and flying objects that are associated with tornadoes can cause severe loss of life.
(2) Waterspouts are not so dangerous because they can be seen and avoiding action taken.
(3) The strong winds that sometimes accompany dust storms can cause the dust to do a great deal of damage, especially to delicate machinery, by penetration and abrasion.

Who are the vanishing peoples of the world?

It is not only animals and plants that are in danger of disappearing from our planet in the face of the industrial 'advanced' nations. There are many races of people that have survived in a very simple, stable way of life for many centuries only to be threatened with destruction in the space of a few decades.

A good example of a vanishing people known to most of us is the case of the Indian of North America. Until Christopher Columbus landed his ship upon the shores of the Caribbean Islands in 1492, the New World and its peoples were virtually unknown, and now less than 500 years later most of the Red Indian tribes are no more, and those that are left have retained little of their culture. Even the surviving tribes have such small populations that many of them may soon disappear. And America is the most powerful industrial nation on Earth.

There are many reasons for the loss of these tribes. The white men came from Europe with the hope of building a new life for themselves in the New World and escaping the poverty of the Old. In most cases, they were welcomed by the Indians. Unfortunately, the white men brought diseases with them. The Indians were not immune and thousands of them died. The coming of the railway opened up the West and supplies of food were required for the rail-

Right:
Many tribes of North American Indians once depended upon the huge herds of roaming bison for their livelihoods. The coming of the railway meant that workers had to be fed and specialist hunters killed millions of these animals. This and loss of their traditional hunting grounds led to the break up of these tribes.

way workers and the new settlers. As a result millions of bison which had provided the Indian with many of his materials, were slaughtered. Slowly, the Indians were driven off their natural hunting grounds on to poorer and poorer lands, and when they tried to fight back, the white man with his army and superior weapons won again.

Above and right:
The sudden coming of industrialization may lead to the destruction of small, self-sufficient communities as people are attracted to factory centres where they may then find themselves living in the terrible squalor of a shanty town.

This is a sad tale, and although it does not happen quite like this today, there are other tribes that are dying out because they cannot adapt quickly enough to the changing conditions which are being forced upon them by industrial man in his search for new lands for building, farming, and natural resources. The Bushmen of South Africa, the Pygmies of central Africa, the Aborigines of Australia, and the Eskimos are just a few of the more well-known races that are dying out or interbreeding with the white men with the loss of their own ways of life. It is thought that there are still tribes of Indians left in the Amazon jungles of South America that have not yet set eyes upon a white face, but these areas are in danger of being opened up.

As is the case with wild life, the loss of these peoples could be thought of as the natural course of evolution, because they are unable to adapt to changing conditions. Generally, however, it is better to have variety of blood, ways of life, and cultures so that there is more chance of man surviving any disaster.

Where does our energy come from?

There are two other questions here which need to be answered at the same time. First of all, what do we mean by energy? And secondly, what kind of energy are we referring to in this question? There is a very exact scientific meaning of the word energy, but for our purposes it is enough to say that energy is the power of doing work. Work can be of all kinds. When you breathe you are doing work and your body is using energy. When you are cleaning the family car, you are working quite hard and using quite a lot of energy. Even when you are sleeping, you are using energy to breathe, digest your food and so on. We get the energy we need for the workings of our body by 'burning' fuel. In other words, we eat food which contains energy locked within it, and as it is digested, the energy is released.

Right:
The long hours of sunshine in tropical zones provides enough energy for a rich growth of plants in which many species of animals can survive.

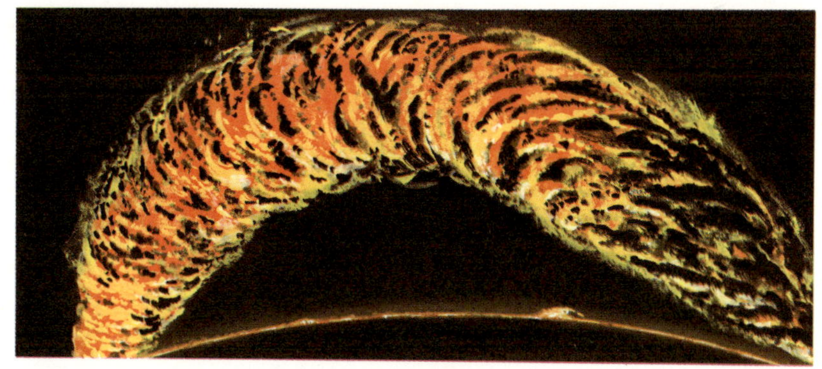

Right:
Originally it is the sun that supplies the energy we use at such an alarming rate. This picture illustrates solar prominences (*see* page 14).

But from where does the energy contained within the food come from? We have already answered this question. The energy is provided by the sun. Plants use the sun's rays to convert carbon dioxide and other materials from the air and soil into proteins. These proteins are then eaten by other animals which we eat together with certain plants. You can see, then, that the energy we use every day comes from the sun through various stages. It is worth bearing in mind that all living things use energy.

Man is rather different to other life forms, however. Have you ever stopped to think how much energy we use in all sorts of ways. Let us list just a few: railways, roads, and ships; electric lighting; all the different forms of heating in industry, in the home, and for hot water; many electrical gadgets. Where does the energy come from? To a great extent we rely upon what are usually referred to as fossil fuels, such as coal and oil. When we burn a piece of coal, in effect, we are releasing energy from the Sun that has remained trapped for millions of years. You have seen how long it takes to lay down the huge reserves of coal, but we are using them up much faster than they can be replaced. The same is true of oil, upon which we rely not only for petrol, but also for many other chemicals. There are also other problems which arise from the widespread use of these fossil fuels. There is the destruction of the countryside in mining for coal and there are all the different forms of pollution that have already been mentioned. Perhaps the next time you switch on an electric light, you could consider the costs to the Earth of our use of electricity on such a vast scale, sometimes for little good purpose.

Above:
Harnessing water power can provide us with electricity but large areas of land often have to be submerged behind a dam to do it.

There are other ways in which we can make use of the energy from the sun. It is possible that in the future we may be able to make greater use of tidal power, wind power, or even solar collectors to convert sunlight directly into electricity. But whatever the source, it is likely that there will be some environmental cost.

How clean is the ocean?

Perhaps it is worth remembering the historic voyages of Thor Heyerdahl in his reed boat, known as the *Ra* expeditions. Thor Heyerdahl had more of a chance to take a closer look at the Atlantic Ocean while he was sailing his fragile craft than anyone in a larger boat would normally have. He was shocked to discover that even in the middle of the Atlantic there was considerable pollution with oil!

The sea has always been thought of as a bottomless pit for all of the waste that man cared to throw into it. In the past, the processes of the sea have been able to cope, carrying waste out and allowing it to decay with the aid of bacteria in the normal way. Now, however, the situation is different. From every country, millions of litres of sewage pour into the sea (some of it via the rivers), some of it treated and some of it not, but this is not the only way in which the sea is made dirty. If you were to go to many of the beaches along the south coast of Britain hoping to lie on golden sands and swim in crystal clear waters, the chances are that you would soon find your clothes caked with sticky black tar-like material and you would be swimming in dirty brown water.

Right:
We have often been tempted to treat the oceans as a giant dustbin. A walk along any beach will convince you how clean the oceans really are.

More recently another problem has arisen. The North Sea Oil discoveries would seem to have been made at a very opportune moment with fuel shortages and rising prices, but they do not come without their price. More and more fishing boats have been finding their nets entangled in debris from the drilling rigs, and the catches of fish are being affected.

Right:
Dispersing an oil slick with detergent can often cause more damage than the oil itself would have done.

Atomic energy seems at first sight to be a much cleaner way of producing the much needed electricity to light our streets, warm our homes, and power our industries, but once again there is a cost to the environment. Unfortunately, nuclear power stations produce materials which cannot be used further for power but which are still radioactive, and which can remain so for hundreds of years. What is to be done with these waste products? One solution has been the bottomless sea again. The wastes are sealed in stainless steel containers and dumped in the ocean. It has been suggested, however, that the containers will have corroded away long before the products are safely non-radioactive.

All the wastes from many factories are transported by the rivers into the sea. Detergents are used to disperse oil slicks, but may even cause more damage than the oil itself. Apart from the thousands of seabirds that die each year as a result of oiling, there may be other effects that we do not know of. It seems that the days of 'sailing the ocean blue' are over.

Right:
Oiled birds die terrible deaths but they are only indicators of other harm that is being caused.

Is fresh air really fresh?

How many times have you heard someone say, 'I think I'll go out for a breath of fresh air'? Have you ever stopped to think what this means in our industrialized world? Of course, if you happen to be living high up among the mountains where there is little or no industry and few people, your chances of finding fresh air are quite good. But what of the built-up areas of Britain and Europe or America? Even if you have not experienced it yourself, we expect you have heard of *smog*. This word is a mixture of the words 'smoke' and 'fog', which is very apt because smog really is smoky fog. London was once famed for its 'pea-soupers', that is, fog containing so much sulphurous waste that it resembled the yellowish green colour of pea soup and was almost as difficult to breathe and see through. These smogs usually occurred during the early months of winter when the fogs that would normally be present at that time of year became contaminated with all the smoke from cars and factory chimneys.

It is interesting to note how much influence pollution of the air, at first by smells of household waste and later by pollution, has influenced the distribution of people in cities like London. As you probably know, London has an 'East end' and a 'West end'. Until quite recently the East end has been by far the poorer half of the city. This is because winds usually blow from west to east carrying the dirty air with them. People who could afford to choose where they wanted to live went to the western side leaving the smelly east for those with smaller incomes. In Los Angeles, in America, which is also famed for its smogs, the prevailing winds are from east to west so that the London situation is reversed.

Above:
Clean Air Acts in many areas have led to fewer scenes like this and smogs are less frequent than they were. However, if we use large quantities of fossil fuels it will be at some cost to our environment elsewhere.

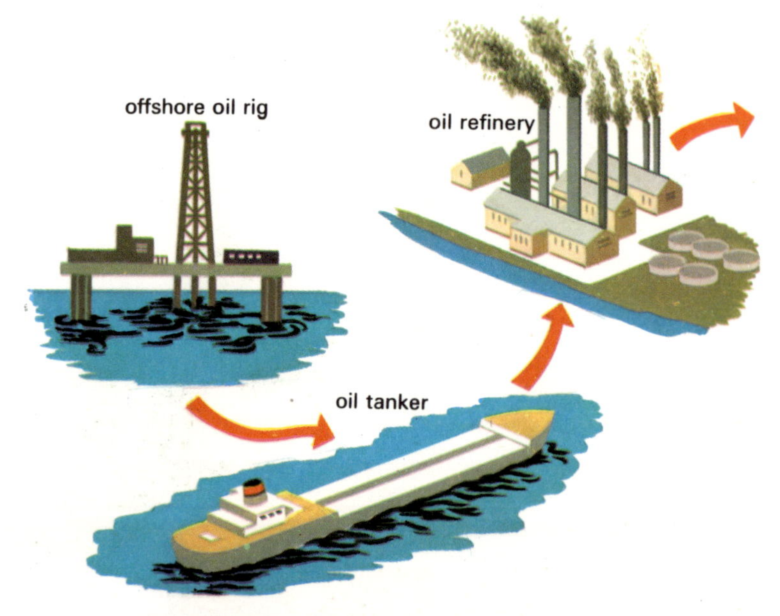

Left:
When no wind blows, smoke may rise higher, but in calm, cold weather the smoke may not disperse and smogs could arise.

Today, many industrial countries have introduced 'Clean Air Acts' preventing the burning of ordinary coal in homes in certain areas and controlling the amount of smoke that factories can release. This means that some areas, such as London, certainly, have air containing much less dust and dirt, and smogs are almost a thing of the past. In general, however, the air in our countries is far from fresh. The millions of cars on the road pour vast quantities of poisonous carbon monoxide gas into the atmosphere. It has even been suggested that if supersonic flight became popular, with the 'planes flying at such high altitudes, the upper atmosphere could be affected in such a way as to permit deadly cosmic rays to reach the Earth.

The next time that you see some plants or trees next to a busy road, take a closer look at the leaves. You will probably find that they are covered with a film of black, oily dust. Remember that this is the same air that we are all breathing.

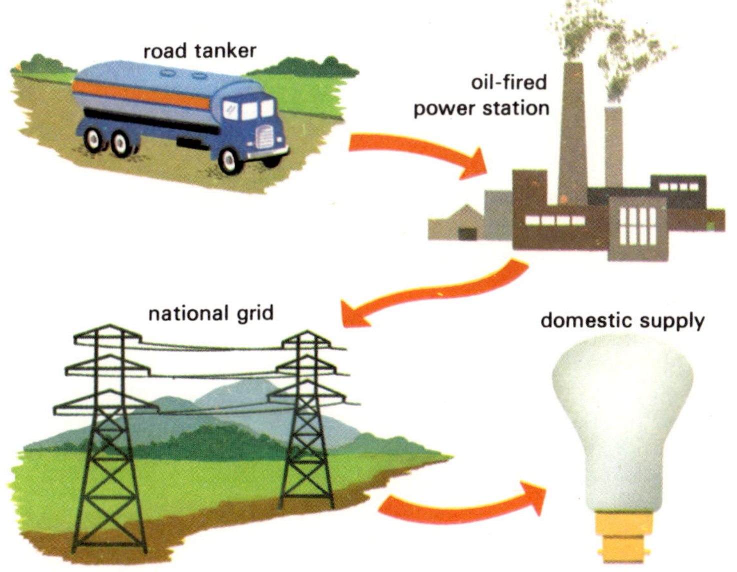

Left:
The next time that you switch on a light try to think of all the processes that have been involved to produce the energy.

Why do we use the land?

Throughout this book, we have looked at many aspects of the Earth, its place in the Universe, its life, its oceans, the upheavals which mould the planet's surface, and so on. Like every other form of life, man has a place in the natural web. Also like most other animals, man tries to change his environment to suit his own needs better, but man has been much more effective in changing the shape of the Earth than any other creature.

Above and right: Thousands of hectares of land are being consumed every day as our villages grow into towns and cities and large areas of good soil are stripped off for certain kinds of mining.

We use the land for all kinds of purposes. First of all, we need the land for food. In the past, man has lived a hunter/gatherer type of existence filling his belly by hunting animals and catching fish, and gathering nuts and berries from the woods in which he lived. He did not always find it easy to ward off the pangs of hunger, particularly in times of drought or following unusually hard winters.

The next step was probably towards some primitive kind of agriculture. Primitive man would clear areas in the forest by slashing and burning so that he could then plant his corn, and when the land was exhausted of its goodness he would move on to another area to begin again. This was the technique used by men of the Bronze Age, and led to a more settled existence. Instead of men finding shelter wherever they happened to be and leading a nomadic life, towns and villages grew up, and as we are only too aware, these small settlements have expanded into the vast industrial cities that most of us live in today. This is a second use to which we put the land. As the population of the world grows and there is a demand for ever-improved living conditions our cities eat up more and more of the countryside. Land that once was thought of as waste now provides us with valuable building plots, and we are even claiming back land from the sea.

Food and shelter are, of course, our main requirements, but these are not the only claims we make upon the Earth's thin skin. Modern living with its electricity, motor car, and industries means that we need to drag many raw materials from the land. We need the fossil fuels for energy so that the land is dotted with oil rigs and slag heaps and pitted with mines and open cast holes. With the coming of the petrol shortages, it has even become an economical proposition to mine the oil shales to extract this valuable commodity. This requires the stripping of hundreds of square kilometres of its vegetation and topsoil, perhaps never to recapture its former beauty.

In addition to all these things we need the land to look at. We have evolved as part of the Earth, and despite urban living we still need to be able to take pleasure in the beauties of the countryside. We use the land for many other forms of recreation such as football and cricket, athletics and even motor racing.

Right:
We need rural areas not only to grow the food we require but also to provide us with areas where we can get away from the stress of urban life for a short time.

What foods do we eat?

In answering this question, it is worth taking into account first of all, that generally speaking man is an omnivore, that is, he does not confine himself to eating just animal flesh like a carnivore, nor does he just eat plant material as does a herbivore. We eat all kinds of foods. The type of food we eat depends upon a number of conditions, such as where we happen to be living, what kind of work we are doing, and even taboos placed upon certain foods by religion or culture. For example if you were living in China a good proportion of your diet might be rice; if you belonged to the Jewish faith you would not be able to eat any part of the pig – in other words no pork sausages or rashers of bacon; if you were a Hindu living in India you would not be able to eat beef because the cow is sacred; some religions prohibit the eating of meat altogether in the belief that it tends to make people more aggressive.

Right:
Small scale farming like this had very little impact on the natural environment.

Right:
Modern methods of farming put millions of hectares of wild land to the plough with the burning of forest and the uprooting of hedges. However, it is now being realized how important hedges are to prevent soil erosion.

Right:
The relative amounts of some of the more important foodstuffs we eat, or feed to our animals:
(1) oats, sugar, fruits, soya beans;
(2) millet and sorghum;
(3) barley;
(4) rice;
(5) maize;
(6) wheat.

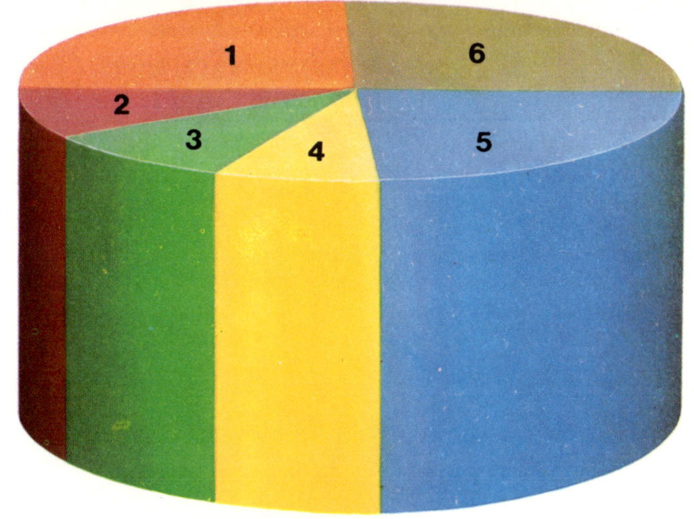

Despite all of these conditions, the world's population depends on a small number of foodstuffs. Most atlases will give you a good indication of what foods we eat and in what amounts. In Britain and Europe people are fortunate to enjoy quite varied diets, but half the people of the world depend on one plant crop for their staple diet – rice.

Rice is a grass which can only be grown in warm climates so that people living in places like India and China rely on it. We need only to include a few other crops such as wheat, maize, sorghum, and barley and we have catered for more than two-thirds of the world's people. With the rapidly increasing world population, crops such as soya beans are being increasingly grown to supplement the proteins provided by meat.

There is also quite a small number of animals that most of the meat eating peoples of the world consume. We expect that you can think of many of them: on your list you might include cows (for milk and butter and cheese as well as for meat), sheep, pigs, and perhaps goats and even deer. (The meat from deer is known as venison.) It has been suggested that as the population rises, with an ever-increasing demand for food, it would be better not to farm animals such as cattle which are not very good at converting grass into protein for us, and it might be preferable to farm animals like rabbits which give more protein per hectare in a shorter time.

Try for a moment to think of your daily intake of food. You might have a cereal for breakfast (probably from wheat) followed by bacon and eggs and toast. For lunch you might have some kind of meat and two vegetables such as peas and potatoes. For tea, a boiled egg and bread and jam might be typical. In addition, there might be some kind of fruit during the day and various drinks such as milk, tea and coffee. All this comes from quite a few plants and animals.

Where do we get the water we drink?

No life on Earth can exist without water. Some animals and plants live in it, others need to drink it either through their roots in the case of plants, or in our case from a glass. But it is not only for drinking that mankind uses water. Try to list all the different ways in which you make use of it. When you get up in the morning, you will wash in water, use water to flush the lavatory, the milk that you pour on to your breakfast cereal is mostly water, there will be water in your breakfast drink, you may boil an egg in water and finally use water to wash up. During the day, you might have many different drinks, use water for cooking, washing your hands, perhaps bathing. You may clean the family car with a considerable amount of water and in summer you may use a hose to keep the garden looking fresh and green. If your family uses a washing-up machine, remember that some types need seventy litres of water to complete the cycle, and of course, there is always the washing machine.

Right:
Water falling on high ground may percolate into the rocks at a lower level to form an aquifer which is under pressure. If the aquifer is tapped, water is forced out under pressure. Wells in London once behaved like this but the water level has fallen now, and this does not occur.

But it is not only in the home that we use water. Industry requires water for all kinds of purposes from cooling to cleaning. Water is also a source of hydrogen.

Where, then, does all this water come from? Of course, in a country such as Britain, this does not seem to be a very sensible question to ask. Britain is an island surrounded by sea and with quite a lot of rain that falls in moderate amounts throughout the year. On the other hand desert areas receive less than 30 centimetres of rain every year, so that if we were living in, say, the Sahara, almost all the available water would be used for drinking purposes.

In Britain and Europe most of the water we use is obtained from the rivers. Of course, as we are all becoming increasingly aware, our rivers today are not very clean, so that the water that comes out of our water taps has been treated very carefully to make sure that it is pure enough to drink. The water then passes through our bodies into the drains and often back into the same rivers from which it was drawn, but a little further downstream. It has been estimated that in a city such as London, drawing water from the River Thames, every cup of water a Londoner drinks has already been drunk seven times before. This does not mean that the water in London is any less clean than further upstream because sewerage works are so good and purifying so effective, but all but the most modern need energy.

Some water can come from other sources, however. As we have already explained, some of the rain falling over the land soaks into the ground to build up a natural reservoir

or aquifer. Again referring to London which lies in a natural basin structure, rain falling on the Chiltern hills and the North Downs percolates into the chalk which forms the bedrock. The level of the water-table at the Chilterns and Downs used to be higher than in London, so that if a well was drilled in London, the water would be under pressure and forced up in an *artesian well*. Now, however the water-table has fallen.

What do we get from the sea?

The first thing to remember, of course, is that the oceans and seas of the world existed long before man was walking the Earth; in fact, long before any other form of life had evolved at all. All life on Earth is believed to have originated from what is commonly called the 'primordial soup'. The first thing that we get from the sea, then, is ourselves in as much as we have evolved from primitive sea-dwelling creatures.

Right:
Salt is very important to us. The word salary comes from the Latin *salarium*, that is, the money that Roman soldiers were given to buy salt. Here salt is being removed from an ancient deposit and from a modern salt pan.

The seas have always been important to man, perhaps of most importance as a source of food of seemingly endless amounts and varieties. When we think of the sea as a larder, the first thing we think of is fish. And every year the sea does yield enormous quantities of different species, but there are many other animals which we eat in large numbers too, such as crabs and lobsters, shrimps and prawns, cockles and mussels, and even octopuses and squid. It has been estimated that in 1968 the world catch of fish and these other marine animals reached a total of almost 60 million tonnes. With improved fishing methods, particularly among the main fishing nations such as Britain and Japan, the catches are still increasing although British fishermen have been having difficulties in maintaining this rich harvest in recent years.

Even the largest of all the animals that live in the sea, the great whales, have long been hunted for meat and oil.

Unfortunately, the increased efficiency of the modern whaling ship has meant that in the last seventy years more whales have been killed than in the previous 400 years, so that two species have already been hunted to extinction and a further five species are seriously threatened. In fact, there was so much call for concern that in 1972 the United Nations called for a ban on all whaling for ten years to give the remaining species a breathing space. Russian and Japanese whaling ships continue to sail, however.

Food is not the only thing we get from the sea. Man needs salt to survive, and of course, the sea is a vast storehouse of it, although in many parts of the world it is easier to mine underground salt deposits. This is particularly true in areas which are not hot enough to cause the sea to evaporate naturally in man-made shallows. The world is now faced with an ever-increasing demand for fresh water. In Israel, for example, and the oil-producing countries of the Middle East, it has become worthwhile to build huge plants to distil fresh water from the sea. It may be that the salts remaining could then be used to provide various elements which are found in the sea.

Sand and gravel are often obtained from the sea floor by dredging ships, and these materials are essential to the building industries. Metals such as copper and manganese may be extracted from nodules that can be found on the ocean floors in some areas. Other metals such as vanadium are concentrated in the blood of animals like the sea cucumber. There is even gold in the sea in far greater quantities than on land although it is very widely dispersed.

Finally, the seas provide us with a place to dump our wastes and highways on which to travel.

Left:
Manganese nodules like these found on the ocean floor contain deposits of important metals including copper and nickel. As copper and nickel are becoming more expensive, it may soon be economic to exploit these riches on the ocean floor.

What makes a desert a desert?

If you were asked to name a typical desert, which one would you choose? Probably the most likely choice would be the great Sahara desert of north-west Africa or the Kalahari desert to the south, or perhaps you might even think of the interior of the Australian continent. What do all these areas have in common? You might be tempted first of all to say that they are very hot. Of course, in the deserts that have been mentioned this is true with temperatures soaring to more than 42°C. In fact, we have it on good authority that in the bush of Western Australia, in the mining communities, it becomes so hot that it is possible to fry an egg on a bare rock and door handles become too hot to hold. Even the main motor route running hundreds of kilometres from north to south, the Great Stuart Highway, is a dirt road because a normal tarmac road would melt. In the houses there, with air conditioning full on, the temperature rarely falls below 22°C, or the average summer temperature in Britain.

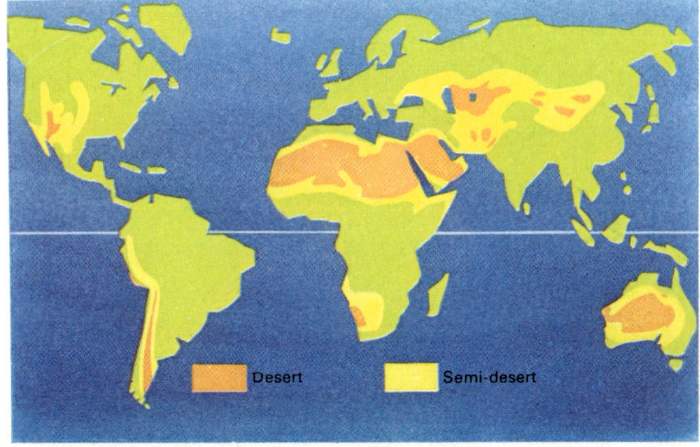

Right:
The map shows the desert and semi-desert areas of the world.

Right:
An oasis in a desert.

Above and right: These illustrations show two aspects of the desert environment.

The next point you might make about these deserts is that they are extremely dry, and it is on rainfall that deserts are defined and not on temperature. A desert is usually defined as an area that receives less than 250 millimetres of rain per year. This is only an average figure. It may be that it may not rain for some years, and then a few years worth of rain may fall all at once. When this happens, temporary rivers form. They may be quite fast-flowing torrents sweeping away the previously dry rocks and eroding quite deep valleys. The waters soon dry up again, however, leaving the valleys dry, the typical *wadis* of desert scenery. Not all deserts are hot. Some are extremely cold, and it is usual to think of the Arctic and Antarctic regions as cold deserts because all of the available water is locked up in the icecaps.

Even in the hottest, driest deserts, not everywhere is dry, dusty, lifeless and strewn with great sand dunes. Rain may fall on the highlands which surround some deserts and percolate into the rocks to form ground water. The sandstones easily absorb the water and an aquifer may develop a few hundred metres beneath the floor of the desert. Where there is an anticline bringing the aquifer close to the surface, or a hollow in the desert taking the desert floor down to meet the aquifer, an *oasis* may develop. This is an area where artesian water reaches the surface. In these areas, the vegetation thrives around the water hole, desert animals use oases to supply them with water to drink, and human settlements may form around them. Sometimes when the rains come in a desert after a number of completely dry years, the water causes seeds that have been lying dormant in the sand to suddenly spring to life and a barren wasteland may appear quite fertile almost over night only to die as soon as the water dries up again.

Where might you expect to find jungles and swamps?

We expect you have read stories of explorers with their native bearers hacking their way through hot, steamy jungles and being plagued by poisonous snakes and spiders, beset by disease and being attacked by savage jungle animals or even head-hunting natives. Perhaps you have read the stories of Tarzan of the Apes by Edgar Rice-Burroughs where a white baby was brought up by a tribe of apes. Most of these stories were set in the jungles of west-central Africa.

Above and right:
A forest like this can support many animals and plants. But there are also tiny communities of animals and plants such as in the human skin (*left inset*) or in a drop of water (*right inset*).

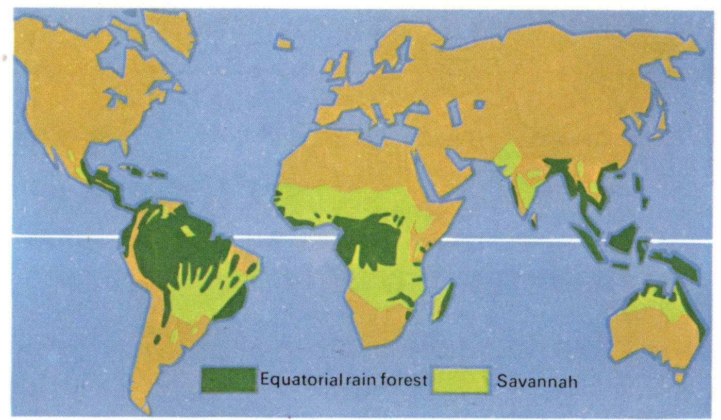

Left:
This map shows the jungle areas of the world.

Below:
A mangrove swamp.

Jungles are to be found in the hot, moist climates of the equatorial belt of the Earth; that is, the region adjacent to the equator. There is probably as much sunshine in these equatorial regions as there is in the hot deserts, but because they are so much damper, the temperatures seldom rise above 35°C. Although, as we have seen, deserts become extremely hot during the day, they can become very cold at night (you will remember the effects that this has upon the rocks). On the other hand, the added moisture in the jungle climates helps to retain the heat built up during the day, and jungles never become very cool. Most Europeans would be able to adapt to the heat of the desert because it is a dry heat, but would soon give way to the steamy heat of the jungles. But it is just this hot moist climate that enables the vegetation to become so dense and provides so many species of birds, reptiles, mammals and insects with such a richness in food.

In some of the great river basins of these equatorial climates, a different type of vegetation may be present. These are the mangrove swamps such as you might find in the delta regions of the Amazon river of South America. The delta itself is formed at the point where the river meets the sea, at the mouth of the river. A delta can only occur where the amount of sediment that is being brought down to the mouth by the river is greater than the amount that can be carried away to sea by the tides and the currents. This means that sediment will be deposited at the mouth of the river.

There are three different types of deltas and two of these are clearly illustrated by the deltas of the Mississippi and the Nile. These muddy swamps may be covered by salt water at high tide and it is under these conditions, along tropical coasts that the mangrove trees thrive.

You can see, then, that in these tropical rainforests the habitat is a very rich one indeed. Tall trees shade the smaller plants, and the area provides homes for animals like the orang-utan and the hippopotamus.

How can you learn more about the Earth?

So far in this book, we have tried to provide answers to some of the questions which perhaps might have been foremost in your mind. But, as we are sure you appreciate by now, there are a great many more questions to ask and very much more to learn. So where do you go from here? It very much depends upon which particular aspects of the Earth interest you most and how much you already know.

In this book, we have concentrated mainly on the geological and geographical sciences. We have touched upon other subjects such as biology and ecology (the study of living communities); it is important to remember that a cockle cannot be separated from the sea in which it lives any more than the Earth can be thought of realistically as being distinct from all the other planets in the solar system. The Universe in which we are a very tiny part indeed, is very complex with millions of related parts none of which is completely isolated from all the others. You cannot study fossils without knowing something about the living animals, nor can you study mines without knowing a great deal about rocks in general.

In any study of any of the aspects of the Earth, there is no substitute for a keen eye and experience, but you can begin in a number of ways and the amateur can sometimes make very important discoveries. There are many other books that you can read dealing with all the natural sciences at every level, but try to ask an experienced person which are the best. Usually, the most recently published books will be the

Right:
Museums can be very useful places to begin your studies of the Earth but there is no substitute for going out and looking in the field for yourself.

Right:
Museums are not only places where good and interesting specimens are displayed but also centres of research. Education plays a vital part in the function of a museum.

most helpful in geology because it is a rapidly changing subject, but older volumes can provide very useful background. Start at your local public library.

As you know, you can see different rocks, minerals and fossils all over the world. It would be very difficult for you to visit all the important localities, but you can learn a great deal by visiting museums that have Earth science exhibitions. Start with your local museum which should have interesting specimens collected in the area and detailed information about your locality. Then try to pay a visit to one of the larger museums such as the Institute of Geological Sciences in London where the displays are constantly being changed and up-dated as more is learned.

If you wish to take a more serious interest in geology, most schools in Britain and throughout the world run courses which are a good basis and will equip you to begin one of the more advanced degrees at university. Most important of all, go out into the countryside and look at it with a critical eye and write down what you have seen.

Right:
The facade of the Geological Society in Piccadilly, London.

What does the future have in store for our planet?

This question has many parts to it. We know, for example, that eventually our star, the Sun, will die and in doing so will destroy the Earth. The very distant future, then, means the end of the Earth, but this will not occur for many millions of years and need not concern us here. It has also been suggested that our climate is changing and that we are at the beginning of another Ice Age, but opinion is divided. It is enough to say that large scale changes to the planet as a whole take place so slowly when compared to our short life span that they may go unnoticed, disguised by the much more rapid changes which man is causing to his own life and that of all other living creatures on Earth.

The usual way of attempting to 'see' into the future is to look at the way changes have occurred in the past up to the present day and then try to project into the future, but this is not easy because it is not always possible to anticipate the effects of a particular set of circumstances. Try a simple experiment. Draw two lines on a piece of paper at right angles to one another to form the axes of a graph. The graph could represent, say, the rate at which a hot water bottle cooled after it was placed in your bed. Suppose one axis represented temperature and the other time. Suppose also that you had measured the temperature on two occasions so that you could plot two points on your graph. These could be joined by a straight line, and if you wanted to know the temperature of the bottle sometime later, you would simply project that straight line to the required point in time. But if, in the meantime, someone removed the blankets from the bed, the bottle would cool much more quickly and your prediction would be wrong.

Right:
Our wildlife communities are very important to us and to guarantee our own future we need to look after them very carefully.

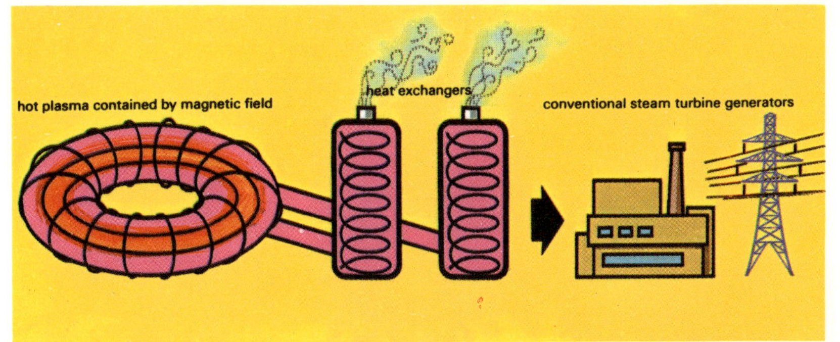

Left:
It has been suggested that the energy released when atoms join together (nuclear fusion) could be used to fuel our power stations. Nuclear power stations today rely on the energy released when atoms are split (nuclear fission).

The subject of many science fiction stories deals with ideas of what conditions might be like in years to come. Some suggestions seem to be wildly far fetched, but a hundred years ago who would have believed in the possibility of supersonic flight or submarines that can stay under the Arctic ice for two years without surfacing, or even in satellites circling Venus? On the other hand in the heyday of the motor car a few years ago, most of us would have laughed at the idea of fuel shortages. How do we know what the world's population might be in a hundred or even twenty years time? Can we predict whether science and technology will be able to find solutions to the problems of overcrowding, food shortages, and pollution choking our rivers and seas, or the ever-decreasing numbers of wild animals and plants with the accompanying dangers to our own species. Will we be able to take a rocket to a distant planet as easily as we can now take a bus to the next town, or will we return to more rural living, each community growing its own food and making clothes by hand? Then there are the dangers of nuclear war. If man is to survive in peace and plenty, the watchword seems to be to try to foresee the consequences of any action rather than wait and see, by which time it may be too late.

Above:
It is possible to cause rain to fall in areas of drought, provided that there are clouds in the sky, by seeding them with dry ice sprayed from an aircraft flying above the cloud cover.

What do you know about the Earth
Index

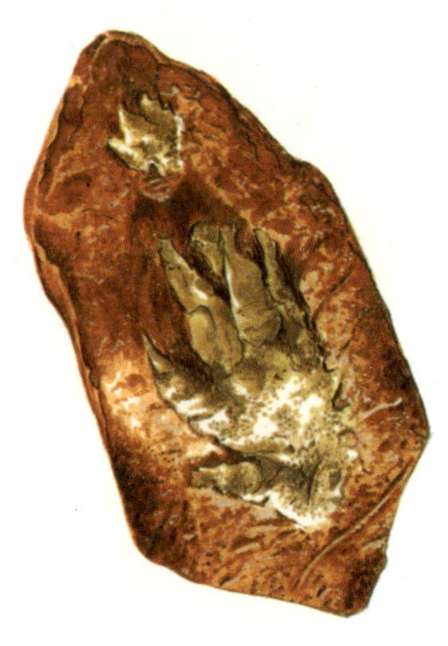

Acknowledgements

The illustrations in this book have been selected from the following Hamlyn all-colour paperbacks:

Aircraft by John W. R. Taylor
Animals in Danger by John Sparks
Astronomy by Iain Nicolson
Beachcombing and Beachcraft by Joyce Pope
Chemistry by John O. E. Clark
Ecology by Michael Allaby
Electricity by D. R. G. Melville
Evolution of Life by Catherine Jarman
Exploring the Planets by Iain Nicolson
Fossil Man by Michael H. Day
Fossils and Fossil Collecting by Roger Hamilton
Geology by A. J. Smith
Physics by Alan Isaacs and Valerie Pitt
Prehistoric Animals by Barry Cox
Rocks, Minerals and Crystals by D. G. A. Whitten and Dr D. C. Almond
The Animal Kingdom by Sali Money
The Earth by I. O. Evans
The Plant Kingdom by Ian Tribe
The Weather Guide by A. G. Forsdyke

Jacket photograph: NERC Copyright. Reproduced by permission of the Director, Institute of Geological Sciences.